Oil and Water

Oil and Water

The Struggle for Georges Bank

by William H. MacLeish

The Atlantic Monthly Press   BOSTON / NEW YORK

FIRST EDITION

Portions of this book have appeared
in *Smithsonian, New England Monthly,
Oceanus,* and *Oceans.*

LIBRARY OF CONGRESS CATALOGING IN PUBLICATION DATA
MacLeish, William H., 1928–
  Oil and water.

  Includes index.
    1. Fisheries—Georges Bank. 2. Offshore petroleum
  industry—Environmental aspects—Georges Bank.
  3. Prospecting—Environmental aspects—Georges Bank.
  4. Petroleum in submerged lands—Georges Bank.
  I. Title.
  SH225.M33   1985   333.91'7   84-72093
  ISBN 0-87113-007-6

                           MV
            Published simultaneously in Canada

            PRINTED IN THE UNITED STATES OF AMERICA

I dedicate this book to
Elizabeth Libbey
my love
and to
the memory of
John Cushman
my friend

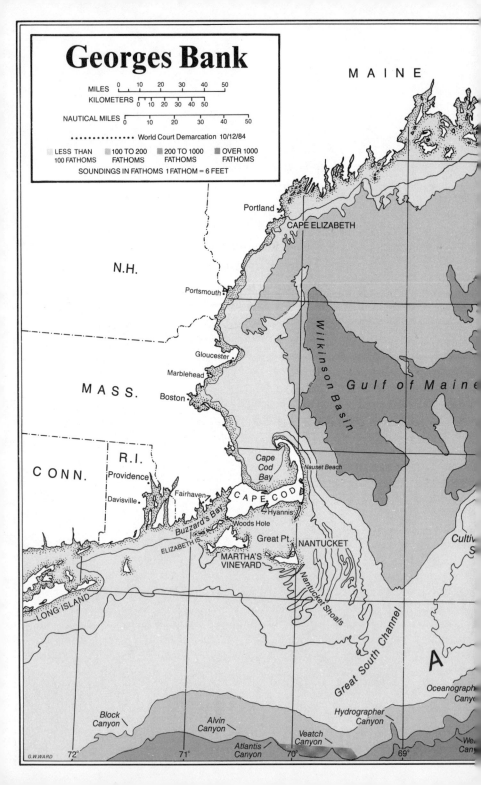

Georges Bank

MILES 0 10 20 30 40 50
KILOMETERS 0 10 20 30 40 50
NAUTICAL MILES 0 10 20 30 40 50

•••••••••••••• World Court Demarcation 10/12/84

LESS THAN 100 FATHOMS | 100 TO 200 FATHOMS | 200 TO 1000 FATHOMS | OVER 1000 FATHOMS
SOUNDINGS IN FATHOMS 1 FATHOM = 6 FEET

MAINE

N.H.

Portland
CAPE ELIZABETH

Portsmouth

Gloucester
Marblehead
Boston

MASS.

Wilkinson Basin

Gulf of Maine

R.I.

CONN.

Providence

Davisville

Fairhaven
CAPE COD

Cape Cod Bay
Nauset Beach

Hyannis
Woods Hole
Buzzard's Bay
ELIZABETH IS.
Great Pt.
NANTUCKET
MARTHA'S VINEYARD

Cultiv
S

LONG ISLAND

Nantucket Shoals

Great South Channel

A

Block Canyon

Alvin Canyon

Atlantis Canyon

Veatch Canyon

Hydrographer Canyon

Oceanograph Canye

We
Can

G.W.WARD

72°   71°   70°   69°

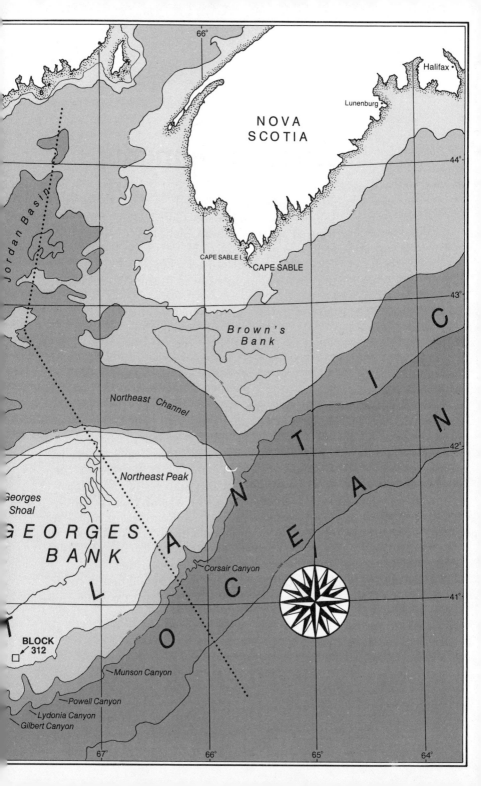

Acknowledgments

THE help I received from the people in these pages is, I hope, self-evident. People like Bob Garvin and Lars Sovik and Harold Moore not only answered my questions but on occasion kept me out of harm's way. People like Bob Graves and Doug Foy showed a degree of accessibility and candor not commonly found in their professions. Thanks be to them.

And to others. First and foremost, to the late W. Van Alan Clark, Jr., of Marion and Woods Hole, Massachusetts, whose generosity enabled me to finish my work. And to Dr. John Steele, director of the Woods Hole Oceanographic Institution, who made it possible for me to begin. To Dr. Richard Backus, editor of the Oceanographic's scholarly study of Georges Bank (MIT Press, 1985), who opened his files to me. To Judi Hampton of Mobil, one of the finest public affairs officers I have dealt with. To Forrest Pierce, owner of Deep Sea Trawlers in Lunenburg, Nova Scotia. To nameless sources at the Ministry of External Affairs in Ottawa and the Departments of State and the Interior in Washington. To Emily Bateson of the Conservation Law Foundation and Pat Murphy of the Massachusetts Office of Coastal Zone Management. And to my sage and salty friend Dana Densmore, who set me straight so many times.

If, despite all this help, there are errors of fact or judgment in this book, the responsibility is entirely mine.

Contents

Introduction

D URING his short and restless tenure as Ronald Reagan's first Secretary of State, General Alexander Haig worried that the world might be entering an "era of resource war." He was talking globally, as secretaries of state like to do, and he was thinking more of the oil of the Middle East than any American resource. Yet Haig's worry, somewhat toned down, can be applied within the United States. Conflicts over resources have been with us pretty much since the first days of the Republic. We pestered each other over land and water and wood and game animals long before we killed each other over slaves. We are more civilized now: we deploy injunctions instead of infantry; we spend years and millions counting legal coups on each other. But the increased humaneness of the attack should not hide the increasing seriousness and importance of the war.

More of us than before now believe — fear, really — that the nature we once thought of or took for granted as abundant is really a nature of limitations. That, given the American character's often obsessive need for optimism, is a cruel dawning, and we have trouble with it. We say there is bound to be more oil here, more fresh air there. We ignore it; we farm land and mine water as if we could go to the store when we needed more. And, increasingly, we lay regional claim to the resources that remain, contesting the national interest in them. We do that most often in the West; regional demands there for a greater say in the exploitation of water and energy resources are described as a rebellion — the Sagebrush Rebellion.

This book examines another resource revolt — not a western shoot-out, though some tough Texans walk these pages, but essentially a Yankee insurrection. The battleground is not ground at all but some twenty thousand square miles of sea well off Cape Cod called Georges Bank. Georges is a ship-killer, a man-killer, and one of the richest fisheries in the world. Long before it was named, it unsettled the bowels of Portuguese or Englishmen — possibly even Vikings — who looked up from their catch to see its breaking shoals to leeward.

The insurrection itself involves two resources: cod and haddock and flounder and scallops and lobster, which have brought fair to fine living to ports in Massachusetts and Maine for upwards of three centuries now; and petroleum, whose existence deep in the sediments under Georges Bank has been thought likely enough by the federal government and several of our most powerful oil companies to warrant the expenditure of hundreds of millions of dollars in a hunt for it.

The insurrectionists are an odd lot. They include some but not all of the offshore fishermen in New England, a remarkably effective collection of fishermen's wives, lawyers working for the commonwealth of Massachusetts and for private groups, and representatives of several environmental organizations. Those at the edge of the movement want no offshore drilling at all, and the arguments they use appear to be based more on intuition than the scientific record. Those at the center say that they are not against oil exploration on Georges Bank — they know the fuel that keeps New England alive through its winters must now be imported and wouldn't mind having a source a bit closer to home. What they oppose is exploration that moves faster than does accurate assessment of what drilling rigs and production platforms may do to the fishery.

Even the loyalists in this battle, those arguing for what they see as the national interest, are not natural allies. Those oil companies willing to gamble on discovering commercially valuable reservoirs under Georges Bank find much at fault with the federal offshore leasing program, which, even under the loose rein of the Reagan administration, constrains their operational options and often forces extremely expensive delays on them. But the companies have no choice; if they want to drill offshore, they must lease their drilling rights from the federal government and follow the federal rules. The government itself, again even the laissez-faire Reagan government, must act under the law as the steward of public lands, including the gravels and heavy sands of Georges Bank, and thus must treat the oil companies as tenants rather than partners.

The government must also act under the law as the manager of the nation's offshore fisheries. As such, it is regarded often with suspicion and at times with fierce hostility by those who trawl, dredge, or set pots on Georges Bank. The fishermen have relented occasionally, as when Washington drastically reduced the fishing rights of foreign fleets in American waters or when it decided to battle the Canadians in court over the ownership of Georges. But always the mood of these captains courageous, these relics of rugged individualism, tends to reflect the weather of their favorite fishing ground — stormy and unpredictable.

This resource rebellion began before I came to it and no one can predict when it will end. Nothing short of the encyclopedic could depict the policies, the issues, the theories of conflict, the cross-purposes generating and generated by the struggle over Georges Bank. But there is another way into what has been a story that has fascinated me for five years, and that is through the people who make the insurrection and those who seek to counter or contain it. The people I spent most time with are hunters and gamblers. They are the Yankee fishing skippers who put their boats, their crews, and themselves at risk to find and take full trips of fish and bring them to auction in New Bedford. They are the Canadian scallopers who, denied access to all of Georges, strip the beds on its tip to rocks and empty shells because they can see no alternative. They are the oil men from Texas and Louisiana who put their rigs, their crews, and themselves at risk to sink remarkably small holes to remarkable depths because their instruments and their hunches tell them they might just find enough petroleum to turn a profit.

People like these make life difficult for the other people in the book, especially the federal managers — from cabinet secretaries on down — who must carry out the will of Congress as translated into a corpulence of regulation and stipulation. In turn, the weight of the federal system — if the word really can apply to a mess of squabbling agencies, each with a prized bit of ocean business to oversee — bears down on the region most concerned with Georges Bank. New England, Massachusetts in particular, has reacted angrily and at times effectively. The state, aided and prodded by young environmental lawyers, has repeatedly gone to court against some of the most powerful bureaucrats in the land. The people who planned and fought in those suits are in these pages.

So are the people in the middle of the fight, the scientists. Ocean research is a relatively new enterprise, and the ocean remains largely unobserved. Yet marine researchers in the great laboratories of Woods Hole and elsewhere are being asked — by legislators, by bureaucrats, by oil companies, by fishermen, by the press — to answer questions often impossible to answer given their present knowledge and, more important, their present ignorance. Many of the impossible questions should more

properly be put to philosophers and other students of social values. That adds to the frustrations of the oceanographers as they struggle to collect and interpret data on
the swirl of water over Georges Bank, the great confusion
of life it supports, and the effects certain accidents and
by-products of drilling might have on certain species
under certain circumstances. In their laboratories ashore
and on research vessels streaming great arrays of instruments over Georges, these men and women conduct
their research and discuss their problems.

I come away from writing this book with admiration for
most of the people in it. I will remember longest those I
met at sea, on the boats, in the planes and helicopters, on
the drilling rig. Theirs is a dignity of dangerous work
done well. From the others I learned something that is
easy to forget in this surly world: that even battles as
heated as the one over Georges Bank can be fought de-
cently and with an eye as much to resolution as to vindi-
cation. Whether there can soon be a resolution to this or
other regional conflicts over resources is a question I
cannot answer. No one I talked to in preparing this book
likes the prospect or actuality of prolonged and repeated
litigation. And yet it is a system that, despite its waste and
crudity, works to keep federalism functioning. It will
have to do, I think, until we can bring ourselves to resolve
the grinding friction between our enterprise, which is
private, and the resources upon which the life of that enterprise depends, so many of which belong to all Americans, and in the case of oceans and atmosphere, to the
world. And it will have to do until we can come up with
and develop trust in a system that will recognize when a
region is being asked to assume risks of resource exploitation in the name of the national good and that will
make timely compensation for damage done, if damage
there be.

William H. MacLeish

Oil and Water

1

First Flight

THE Coast Guard air operations are hidden away in Otis Air National Guard Base, a great stretch of upper Cape Cod flats full of ghost barracks and barricaded roads. A Massachusetts air unit uses a bit of it to send its training jets howling up in paired takeoffs. Right next to that commotion sit four Goats — Grumman Albatrosses, built a few years after World War Two. They invite the eye — hulls flared in the bows like an old sports fisherman you might see in a Bogart movie; slab sides; high wings. They should be riding to anchor somewhere in the Caribbean, lifting off the Amazon into a red sky.

The flight board in the control room lists Coast Guard patrol 7250, takeoff at 0800, pilots Sultze and Post. Sultze is already in and on the phone, joking with a friend about the load of extra people he's taking out to Georges Bank. Who? A turtle watcher from the University of Rhode Island; a law-enforcement man from the National Marine Fisheries Service, simply called Fisheries or, sometimes, Nimphiss, in the port towns here, the manager of what swims and crawls in federal waters out beyond the three-mile state limit; some writer from Woods Hole — probably from the Oceanographic or one of those ocean science labs down there; and one of our own Coasties, a chief warrant officer from Brooklyn. Flying heavy today, he says, and we may have to kite all over because of the fog.

The room fills with Coasties, men and a few women, most looking enviably at peace with themselves and the early summer morning. Peace, this small way into a day of flight, is for the young. Somebody invites somebody else over that evening for hamburgers. Better warn your wife, says the invitee, or she'll cut your throat. She will anyway, says the inviter.

Pilot Post arrives and talks with Pilot Sultze, who is taller by a head and hides his youth behind a moustache. Post wants to know if the patrol may be diverted to search and rescue, SAR, for a sloop that sank during the night near Georges Bank. No, Sultze says, we're scheduled to chug up into the Gulf of Maine, swing out to Georges to visit a piece of ocean Fisheries has closed to the boats so that the stocks can get a rest. Then home.

The chief warrant officer from Brooklyn is just passing through and wants to take another ride on an old Goat while they're still in service. He slouches in a chair while others of his revered rank razz him about last night. All that bourbon you drank, you must have a mouth that would gag a maggot. No, he says, "bourbon works on the brain. If you don't have too big of a brain, there's not too much working space for the bourbon." Sultze heads for the door. The chief rises and, with dignity, shoulders his hangover and follows the pilot out to the flight line. We struggle into survival vests loaded with flares and dyes and signalling devices. The flight mechanic says that all windows *will* be kept open during flight. Otherwise he will get sick, and we don't want *that*, do we?

Sultze is on the intercom. Our primary mission, he says, is to check the position and registration numbers and gear type of as many fishing boats as we can. If we spot one doing what he shouldn't be doing, he'll be reported. That can mean a fine or confiscation of gear. Fishing in the closed area we'll visit is one violation. Another, though the pilot doesn't mention it, is heaving to and taking on a load of marijuana or a packet or two of cocaine. Maybe a couple of hot parrots from a South American jungle — they bring a lot from those who don't want the inconvenience of observing American quarantine laws. Now that the federal pressure is high down in Florida, the dopers are moving north toward the blind bays and inlets and salt creeks along New England and the Canadian Maritimes.

The turtle man sits next to me. He makes these flights regularly, looking for whales and other mammals and for his turtles. Leatherbacks and loggerheads come north following their favorite jellyfish. Some have been spotted as far north as Baffin Island, he says. The study he is working on covers most of the western North Atlantic, the great reach of sea that bathes and beats on much of the American and all of the Canadian eastern shore. Its purpose is to take inventory of these creatures, to have some sort of a record so the study makers and their colleagues can measure the effects of certain activities. What activities? Well, things like marine pollution or like the oil drilling that appears to be in the works for Georges. The Department of the Interior has designated the Bank as one of many likely reservoirs it thinks may lie offshore. It wants the oil to increase domestic supplies, reduce dependence on untrustworthy foreigners, improve the economy, and in general make the Republic shine. The oil companies want the oil for all those reasons plus profit.

Sultze and Post complain to each other and to the tower. The engines are not about to synchronize; we can hear the flutter of their dissonance slow down and speed up as the pilots fiddle with this and that. These Albatrosses are well past their allotted time on earth, rebuilt, coaxed along. They'll be replaced soon by jets, though the turtle man isn't sure he understands why anyone would want to use a jet when the job calls for pooping along and having a good, long look. The chief from Brooklyn leans over to tell us our airplane is too old to land on water safely any more. The hull is too fragile. Perhaps he has more of a hangover than he admits to. Perhaps not. The engines are still in rebellion, so we crawl back to the flight line and clamber out.

The turtle man says this sort of mission scrubbing is normal. "If you're going to spend fifteen, twenty hours a week in the air, you want to be sure you're in something that won't fall down."

"Hell," says the chief from Brooklyn, "these are good airplanes."

"Until you try to start 'em."

"Naw," says the chief. "These things never break in the air. They may break when you land or take off, but never in the air."

The pilots tell us next that we can use another Goat if the mechanics can fix its fuel leak and faulty carburetor. They do, and yet again the engines play games, but this time Sultze and Post win out. We lift easily, banking away to the sea.

The kettle-hole ponds and shingled condominiums dwindle. For a second, the Albatross holds a course that would take it east, parallel with the Cape's southern beaches, and out to sea over Great Round Shoal Channel, over the northernmost dunes of a subsea Sahara, the Nantucket Shoals, to the Great South Channel that cuts through the sands to connect the southern part of the Gulf of Maine with the Atlantic.

There, just on the other side of the channel, a hundred statute miles or so east-southeast of Nantucket Island, is the base of a giant, submerged thumb stretching away to-ward Nova Scotia. *Georges Bank.* Ten to twenty thousand square miles of it, depending on which depth line or iso-bath you choose, wild land under a wild sea. Serried shoals run across it, some only a few meters under the surface of the water. Deep canyons slice its outer edge.

Georges is only one of several shallowings along the Atlantic continental shelf of North America. It is much smaller in mass and myth than the Grand Banks further up off Newfoundland. It has no captains courageous, and its shanties are unexceptional: "Come all you brave undaunted ones / Who brave the winter cold. / And you that sail on Georges Bank / Where thousands have been lost." But it is, acre for acre, one of the richest fishing grounds there is, a roil of biological productivity in a world ocean that tends to be only patchily productive. That is why the turtle man is on board, why the enforcement man from Fisheries, who looks the way Wyatt Earp should have looked, sits by his window with his notepad clamped low on his thigh.

We turn north, drop down over the dunes of Provincetown at the finger of the Cape. Within a few minutes, we are "rigging" boats. That means, in Coastese, that we are identifying them and noting their location. All eyes are peeled for registration numbers and names, for hulls riding suspiciously deep in the water, boats suspiciously overcrowded, suspiciously scuffed on the hull from transferring suspicious cargoes in a working sea.

"Ten seconds, on our right," says Sultze. A scalloper comes under the wing, close to a hundred feet long. One massive metal dredge has just swung inboard and dropped to the deck. The haul is small, mostly rocks from the bottom, but we can see the tan and white of fair-size scallop shells among them. The deckhands, seen from this height, are all ass and elbow as they scrabble through the haul throwing the coins of their realm into wire baskets and hauling them away for shucking.

A stern trawler snaps into sight under the starboard wing with what is left of a small spruce bound to the masthead. The law-enforcement man says that's the badge of the boats with the biggest catches, the chiefs among the fishers. Another stern trawler, pretty far to port. The law-enforcement man gets the name and num-

ber from the flight engineer. "Think that's out of Fairha-
ven," he says. "Norwegians run her. Finest kind."
("Finest kind." That's coastal Yankee for first rate, great
stuff.) Here comes a name: *Divine Creator*. The Portuguese in New Bedford and the Italians in Gloucester call
their boats after their wives or the saints. This might be a
Moonie boat. The Moonies, Americans who follow the
Reverend Moon of South Korea and his controversial
Unification Church, came to Gloucester a few years ago.
Now they operate some fishing boats and shore facilities,
and that outrages a lot of fishermen who resent what
they see as excessive low-wage competition.

A side trawler appears directly beneath us. She is old,
wood-hulled, and she sets her net in the old way, over the
starboard side. Her pilot house is aft, and she carries a
fragile dory upside down like a campaign cap on its roof.
A half-mile to her starboard is a brand-new, million-dollar
marvel of a stern trawler with her net coned behind her,
just beneath the surface, while crewmen in yellow slickers shackle on the trawl doors.

We stand on our starboard wing to get over above a
lobsterman heading out for one of the underwater canyons, Oceanographer Canyon possibly, or Lydonia. His
long stern scoots along under stacks of offshore lobster
pots big as coffee tables that will be set in strings a mile
or more long, marked by buoys.

"Nav, aft." That's the turtle man asking the navigator
for a fix.

"What did you see?" Sultze asks.

"Whale shark. About twenty-five feet long."

"Well, next time let us in on it."

The navigator gives the turtle man a position. Fog fingers stretch toward us from the horizon, and the navigator's radar becomes more important. He spots a target
on it ten miles out in front.

"Put it on the nose, please," Sultze says. He banks the
plane right until he hears the navigator tell him he's on

course for interception and then drops the port wing level again. Silence. Then, "On your right side, fifteen seconds."

As usual, everybody looking to starboard tries to get first identification. No problem, this time. The Canadian maple leaf glistens in the sun. We're about as far north as we're going to go on this patrol, close to Northeast Peak out on the tip of Georges. "Lots of Canadians around now," says the law-enforcement man. "Those scallops down there mean a lot to 'em."

The Canadians claim the northeastern third or more of the Bank, the first joint of the thumb, and the United States claims all of it. The World Court will decide who owns what at some point in the next few years. "Whichever way it goes," says the enforcement man, "somebody will be screaming, our fishermen or theirs, our politicians or theirs."

We swing off the edge of Georges into deep water. A big, blocky Japanese longliner stands by her set, miles of line carrying thousands of hooks baited for swordfish. The turtle man says the number of foreign ships — by which he means those that have to cross big water to get here, not the Canadians — is way down now. "Before the Magnuson Act, before we went to the two-hundred-mile limit in 1977, it was ship city out here. Russians, Germans, Poles, Spaniards, everybody. At night, it looked like Broadway had floated loose."

Sultze declares hunger. He wants mustard on his sandwich, but the flight mechanic sorts through the big ice chest and can't find any. Then the flight mechanic sits down and commences to move through four box lunches all by himself. Those who remember why he insisted on open windows roll their eyes and pray.

We pass a ketch and swing around for another look and a few pictures. The yacht rides high and carries only three: a fat man and a thin woman and a dog with its own square of artificial turf. That's all the rage. After your pet relieves himself, you tow the turf in the wake and then haul it back, squeaky clean, for another go.

"Humpbacks," yells the turtle man. A spout out beyond the wingtip, and then a whale breaches right below us, after a school of herring. "Busy day," chortles the turtle man. Just an hour before, he had spotted — again without letting us in on it — a school of white-sided dolphins. He wants to rig the whale under us. "Will you take a racetrack turn to the left?" Sultze obliges and asks him about humpbacks, whether the plane scares them. Don't know about that, the turtle man says, but he once did, literally, scare the shit out of a whale. Boom, just like that. Must have been dozing on the surface and couldn't face reality when the plane woke him up. The humpback we're turning for has better ears. Halfway through our maneuver, he sounds.

Now the loran is out. Sultze doesn't seem too bothered. He asks the navigator if, in his high-tech training, he ever learned to take a sun line.

"Nope."

"Don't you know where the sun is?"

"In the sky."

The fog is getting thicker and steadier, and it's harder to rig. Someone spots a southern boat, from Texas. A few come up here still, even though scalloping isn't anything like what it used to be on Georges.

We're about over the closed area we're supposed to check. Sultze tells us to keep an eye out for boats trawling the bottom inside the no-fishing line. Two gill-netters are setting out, the monofilament mesh all but invisible in the water. Near them are two side-trawlers, suspiciously close to being in illegal water. We rig them repeatedly, taking pictures. They look small to be out here, more than 170 miles from home.

We head back, scurrying before a tail wind. Someone, looking at the small boats and the big loneliness of the sea, wonders what it's going to be like when there's an oil field down there. No one answers. Something in the silence shows the difference between attitudes toward drilling offshore in New England and along the warm coasts of the Gulf of Mexico. In the South, rigs and platforms have been part of the culture since just after World War Two, when the drillers began following the oil patch south across the salt marshes and on out to sea. Here, where the past has a way of reining in the present, most New Englanders and most visiting tourists think of the sea in terms of the half-mythic men wrapped in oilskins who risk everything for cod and the glory of hunting alone. There, the western North Atlantic is regarded as just another frontier, just another promising spot on the vast continental shelves of the world still virgin to the drill. Here, the offshore banks are the foundations of civilization. We need the oil, says Boston, but wait a minute, let's make sure we get it out safely. By the time you make up your mind, we'll have pumped the Gulf of Mexico dry, says Houston; it's time for you other coasts to lend a hand with the exploration. We were here first, says Boston, we started this country. Why don't you bastards go freeze in the dark, says Houston. See you in court, says Boston. If you've got the gas to get there, says Houston.

At the forearm of the Cape, the sea sands rise under us, blue to green to gold to the dun cliffs of Nauset Beach. We come down steeply. In the classic and apocryphal Goat approach, the pilot is supposed to jam the stick forward until he hears the copilot scream "Jesus Christ" and then pull up, but there is nothing of that sort coming in on the intercom. Just some bitching about the engines acting up again. We climb down and stand on the flight line and shake hands and go our way. The law-enforcement man stays behind to check his positions and see if those trawlers really were over the line.

2

"Let Not the Meanness of the Word Fish Distaste You"

Henry Bryant Bigelow was an outdoorsman of passion and competence. He hunted and fished wherever he could, but his favorite range, at least in his early years, was the Canadian Maritimes. In 1906, he and his wife took their wedding trip on a wild river in Newfoundland. When the canoe capsized and she lost her sneakers, he shot a caribou and made moccasins for her from its hide. He went to sea whenever he could, despite a considerable handicap: he got repeatedly and miserably seasick. At one of his first weekly staff meetings as director of the Woods Hole Oceanographic (he became the institution's first head when he was well along in his career), Bigelow gave a talk on "Seas I Have Vomited In."

One of those seas was the tropical Pacific, where Bigelow, then a newly certified zoologist, had gone with Alexander Agassiz. Bigelow published his research findings on the region's siphonophores, those lovely and often transparent polyps that drift where the sun lights the water. The work won him a good deal of attention, but by then Bigelow was thinking ecological thoughts about a patch of ocean just a day's sail from Boston. In a short autobiography, he described how the eminent Scottish oceanographer Sir John Murray prodded him from thought to action. After the old man had listened to the younger's requests for advice, the following discourse occurred:

Murray: "How much is known about the Gulf of Maine."

Bigelow: "Practically nothing."

Murray: "Can you row?"

Bigelow: "Yes sir."

Murray: "Can you borrow a deep-sea thermometer?"

Bigelow: "Yes sir."

Murray: "Can your wife make some tow nets out of her old bobbinet window curtains?"

Bigelow: "Yes sir."

Murray: "Don't ask me any more damn fool questions."

Ship time — days aboard a research vessel — is to an oceanographer what flight time is to a pilot. Bigelow got his aboard the schooner *Grampus* and other vessels belonging to the U.S. Fish Commission, forefather of today's National Marine Fisheries Service. From 1912 to 1929, he and his colleagues occupied 350 stations in the Gulf of Maine, a good many on its seaward rim, Georges Bank. Occupying a station is roughly the same as setting up a field experiment on land, though usually many times more difficult. At a precisely determined spot in the ocean, the research vessel anchors, heaves to, or, if equipped for it, maintains position with computer-controlled bow-thrusters and other novel propellers. The scientific party then makes its measurements. In Bigelow's case, temperatures and salinities at various depths were recorded in every season of the year, thousands of tow-net hauls were made, and hundreds of bottles were set adrift to map the surface current systems. Fishermen by the dozens were hailed and asked about catches or

the lack of them. "As a result," Bigelow wrote with more
than a little modesty, "it is generally conceded that
oceanographically the Gulf of Maine is better known than
is any other comparable area of the ocean, the surveys of
which have been carried out by a single agency."

To be more accurate about it, the still fairly youthful
Bigelow and his handful of collaborators delivered to science and the public three classic, encyclopedic monographs on the fishes, plankton, and hydrography (the
living resources and their medium) of the Gulf of Maine
and Georges Bank. It was a monumental undertaking,
even for someone with a serene stomach. And it was only
one of several in what Bigelow called a long and active
life. He died in 1967 at the age of eighty-nine. His scientific publications spanned more than six decades, and he
worked at Harvard longer than any faculty member before him. Reflecting on that record, he was moved to ask
a friend on the Harvard Corporation if such service might
not merit some token of recognition, say, a bottle of bourbon. It did. The bottle arrived shortly "with the compli-
ments of the President and Fellows." A few years later,
when Bigelow lay dying in the hospital, some old ship-
mates from the Oceanographic came to visit him. Bigelow roused up out of a shallow sleep and eyed them.
"Why," he whispered, "aren't you people at sea?"

It isn't always easy to find a copy of Bigelow's works
around Woods Hole. Some scientist or other usually has
them checked out — especially *Fishes of the Gulf of
Maine* by Bigelow and his longtime friend, the late William C. Schroeder. But if one comes to Woods Hole for
the fisheries science of Georges Bank, one starts with this
book. It is written, praise be, so all can understand.

I picked up my copy from the bunk of a biologist while
waiting my turn to dive in the Oceanographic's research
submarine *Alvin* to the floor of a deep canyon on the seaward edge of Georges. As we lay to, the little sub edging
across the seafloor hundreds of feet below us, I sat in the

summer sun and read, of herrings and mackerel, of re-
quiem sharks and basking sharks and gurry sharks and
bramble sharks, of witch flounders and hogchokers,
wolffishes and toadfishes, wrymouths, rocklings, dabs,
scads, barn-door skates, and squirrel hakes. And of cod,
haddock, yellowtail flounder, the prizes — along with the
golden invertebrates, scallops and lobsters — of Georges
Bank.

Bigelow chose the haddock as a model to explain his
keys to identification, used much as one would use the
keys in *A Field Guide to the Birds*. If your fish has long
jaws and pectoral fins (those just back of the gills), if it
has three clearly defined dorsal or back fins and two anal
fins, and if it has a dark lateral line running high on each
side and a dark blotch on each shoulder, why, what you
have in your hands is most likely a haddock. The dark
blotch is called God's mark, or Saint Peter's or the
Devil's, depending on your source. Bigelow inclined toward the Devil.

The color of the haddock, Bigelow and Schroeder
wrote, is purplish gray on the back, paling on the sides
"to a beautiful silvery gray with pinkish reflections." The
biggest on record are well over thirty pounds, but most
catches are of fish running five pounds or less. The mature fish seem to prefer depths of between twenty-five
and seventy fathoms, a fathom being six feet. Once they
take to bottom, once they swim down from the bright
surficial waters where they hatched and spent their larval
days, the haddock often seek out broken ground or
smooth areas between rocky areas. They aren't as common as their cousins the cod over rocky ledges or kelp
patches, or as their cousins the hakes over soft, oozy
mud. But give them gravels, pebbles, sands, and clays,
and they will root and feed like sea pigs. They will take
just about any kind of invertebrate — crustacea, such as
small crabs and shrimp and amphipods (relatives of the
sand flea) — as well as small clams and other mollusks,

starfish, sea urchins, sand dollars, brittle stars, and sea cucumbers. And worms: haddock are often found to be packed with worm tubes in areas of Georges where so many of these burrowing organisms live that their habitats are called "spaghetti bottoms." Lots of worms for the haddock, but few fish.

Haddock spawn on and around Georges Bank generally from February through May. A large, sexually aware female may contain 1.9 million eggs or more, each about a millimeter in diameter. Like those of the cod, which it closely resembles at first, and like those of several other commercially important species, the haddock egg is pelagic, part of the plankton that is at once so varied and so vulnerable to predator and, on occasion, to pollution. The fish that take to bottom are hemmed in on the north by the deep Northeast, or Fundian, Channel — they avoid water over a hundred fathoms deep — and that may help to explain their concentration on the Bank. "This indeed," wrote Bigelow, "is perhaps the greatest haddock ground for its size in the world."

Bigelow didn't dwell on the delicacy of the haddock, its supremacy as a chowder fish, the nutty flavor of the smoked flesh — what the Scots, honoring their old fishing port of Findhorn, call finnan haddie. But true to his passion, Bigelow finished his findings with sport in mind. "A haddock of fair size is likely to prove an astonishment to anyone who is lucky enough to hook one while fishing with a light sinker."

The yellowtail flounder, said Bigelow, is right-handed and small-mouthed. Its eyes are on the right side of the flat body and its innards are to the right if you look down on one swimming away from you. It is wide and thin-bodied and of middling size for a flatfish, running around sixteen inches on the average and a pound or so in live weight. Its back — really its right side — is brownish or slaty olive and covered with rusty spots. The name comes from the tail, which is yellowish, as is the peduncle, or stalk of muscle connecting tail and body. On Georges and in the Gulf of Maine, the yellowtail is caught mostly at depths between forty fathoms and sixty.

Like cod and haddock, the yellowtail ranges both sides of the North Atlantic. On our side, it is found from the Gulf of St. Lawrence to the Chesapeake. It is plentiful on the western side of Georges and on the Nantucket grounds near the base of the great thumb.

Bottoms of sand or sand and mud are yellowtail favorites. There it can hide most easily and feed at will on little shrimp and amphipods and worms it takes special interest in. It is no great wanderer. Its thin body makes it less valuable than fleshier flatfish on the market, but its numbers and the ease with which it can be trawled make it a prime target for the net. Yellowtails, said Bigelow, "live rather too deep to be of any interest to anglers."

The cod is the eponym of the gadoids, the family that includes the haddock, the dark handsome pollock, the whiting, and other hakes. It is the largest gadoid. One that must have gone 180 pounds live came in on a fishing schooner back in 1838, and another six feet long and 200 pounds was taken on a line off Massachusetts in 1895. You don't hear of giants now, and that's an indication of a stock under pressure. Thirty pounds is a fine fish these days.

Cod are the nomads of the groundfish. They come to the surface after sardines or other prey. They can live comfortably at 250 fathoms and probably well below that. Young fry come in close to shore, while others take to ground on the offshore banks. Populations of cod travel to avoid water warmer than fifty degrees Fahrenheit, to escape sharks, dogfish, pollock, and other enemies, and to find food. They swim far north, sometimes stranding themselves chasing the smeltlike capelin onto the

beaches of Labrador. Others move far south of New York
or plain stay where they are. In Europe, they range from
the northern capes of Scandinavia to the Bay of Biscay,
and over here from the high latitudes of Greenland almost to Hatteras. "We fancy," Bigelow wrote, "there is
no patch of hard bottom, rock, gravel, or sand with broken shells, from Cape Sable [Nova Scotia] in the east to
Cape Cod on the west, but supports more or less cod at
one time or another." Georges Bank has been the richest
American cod ground, particularly its eastern half. Acre
for acre, it outproduces the Grand Banks.

The cod carries the amazed look of many fishes, but it
carries it regally. Its mouth is wide and its nose is blunt
and its head is huge — often a quarter of its length. The
body is deep and thick with a promise of meat that makes
the hungry hungrier. The barbel, the chin whisker of
flesh, is more pronounced in the cod than in most of its
cousins. The lateral line is pale, and the Devil has not
touched the shoulder. In coloration, Bigelow saw two
groups, the gray-green and the red. Both are thickly
speckled on top and sides with "small, round, vague-
edged spots," brownish or reddish depending on the
ground color of the clan.

The prolific haddock runs a poor second to the cod
when it comes to production. If you were lucky enough to
find a seventy-five-pound ripe female, you would be looking at a factory with a capacity of nine million eggs. One
million is more like it around the Gulf of Maine and
Georges. Cod like to spawn east of angry Georges Shoals
in twenty-five or thirty-five fathoms of water. Other favored spots, Bigelow found, are Nantucket Shoals and
along the twenty-fathom line off Massachusetts and
Maine in the gulf. On Georges, cod usually spawn from
February through April. The tiniest fraction of the brood
survives the first year. Those that live nine years, the
survivors of survivors, might reach four feet.

Gadus is a glutton. The larval cod, like most ichthyoplankton — fish in their first stages — relies mostly on copepods, microscopic crustacea with oarlike legs. And the same diet, varied with amphipods and other small crustacea, barnacle larvae, small worms, and such, satisfies them when they first swim down to the bottom. But before long, Bigelow wrote, "they consume invertebrates in enormous amounts . . . their stomachs are mines of information for students of mollusks." Shells of large sea clams are found nested in cod stomachs, along with those of cockles and sea mussels, all swallowed whole. Add shrimp and crab and brittle stars and sea urchins. Haddock may be catholic eaters, but they do not swallow large shells. Nor do they pursue fish anywhere near so assiduously. Cod take squid, herring, sand launces, capelin, young haddock, mackerel, menhaden, cod. The larger ones take ducks, even old boots. "And they often swallow stones," Bigelow wrote, "but probably for the anemones, hydroids and other animals growing thereon, and not to take on ballast for a journey, as the old story has it."

"Puritan Massachusetts," wrote the Yankee historian Samuel Eliot Morison, "derived her ideals from a sacred book; her wealth and power from the sacred cod." Massachusetts knew that early on. In March of 1784, its legislature voted to "hang up the representation of a cod-fish in the room where the House sit, as a memorial to the importance of the cod-fishery to the Commonwealth." The memorial remains, stirring a bit in the winds of oratory.

Americans today have difficulty conceiving of the cod's importance to their beginnings. After all, we have turned landsmen, going from farm to factory and now to — what? Even if something does call us back to the coasts (almost half of us live within fifty miles or so of big water, and four out of five of us are supposed to do so at the close of the twentieth century), we no longer appreciate fish much. Cancer warnings and the mania to reduce have sent our intake of seafood up over fifteen pounds a year. But we still eat more than 130 pounds of meat, mostly beef and pork, per person per year, and our bellies tend to hang over our belts in testimony to that misproportion.

Harold A. Innis, who wrote a seminal history of the cod fishery in North America (*The Cod Fisheries: The History of an International Economy,* University of Toronto Press, 1978), says the fish had a different impact on his native Canada than on the United States. "While New England, with a more favorable climate and an extensive hinterland, became less dependent on the fishing industry," he wrote, "Newfoundland grew increasingly dependent on it. But throughout the whole [western North Atlantic] the exploitation of the fisheries has been of dominant importance. No other industry has engaged the activities of any people in North America over such a long period of time and in such restricted areas."

Cod was the convenience food of Europe. Properly cured, it could keep well in larder or saddlebag. During Lent, it was a staple, though one that needed a fairly direct approach in preparation. "It behoves," wrote a Betty Crocker of the fourteenth century, "to beat it with a wooden hammer for a full hour and then set it up to soak in warm water for two hours or more." Demand was such that when John Cabot returned from his first voyage of discovery at the turn of the fifteenth century, the accounts that seemed most to warm entrepreneurial hearts were those of fish teeming off the New Found Lands. They could be taken, said one report, "not only with the net but in baskets let down with a stone."

Though there was plenty of activity inshore, the great banks, the Grand Banks of Newfoundland, 40,000 square miles of them, attracted concentrated effort. The British fished there, and Spanish and Portuguese and, in time, the American colonists. But the most persistent were the French. Before the British drove them from Atlantic Canada in the middle of the eighteenth century, the French had built a fleet of more than 500 ships manned by over 27,000 men — far more than you'll find today in the fishing ports of New England.

The Banks fishing gradually ranged out until it covered a chain of shoalings from Flemish Cap to Georges, 1,100 miles down the coast. "They were regarded," an American historian wrote a century ago, "as one of the greatest sources of wealth then known to the world." Some of the catch went green to Europe, pickled in brine. But the best way to preserve it was to go ashore and dry it in the sun. Men killed and governments threatened war over which stretch of beach would be used by what ships.

A hundred years after Cabot, Bartholomew Gosnold sailed into New England waters in search of sassafras, which was thought effective against syphilis, among other things. He found "better fishing and in as great plentie as in Newfoundland." In such plenty that he called a hook of land he encountered Cape Cod. The fish were bigger and fatter than northern stock, and captains later told of "pestering" their vessels with catches too bountiful for the hold. That great booster John Smith came over to look and reported to his sponsors, "Is it not pretty sport to pull up two pence, six pence, and twelve pence as fast as you can haul and veer a line?" And: "Therefore, honorable and worthy gentlemen, let not the meanness of the word fish distaste you, for it will afford as good gold as the mines of Guiana or Timbuctoo, with less hazard and charge and more certainty and facility, and so I humbly rest."

Ah, the early abundance. All you wanted you could get from a small boat. Salmon crowded the rivers, and sturgeon so large they were considered dangerous to the passage of canoes. Everything had its uses: spines from the dogfish for toothache, cod bones to control "women's overflowing courses," sounds or bladders from the sturgeon "which, melted in the mouth, [are] excellent to seal letters."

The sturgeon dwindled. By 1759, fishermen were complaining that sawdust from the mills on the Piscataquis River in Maine was driving away the salmon. Some port towns passed laws barring the taking of some fishes in spawning season and throwing gurry overboard on the grounds, but the virgin yields inshore could not stand the pressure from the world's most unrelenting predator. Old men remembering took to voicing what has become the American lament: you should have seen what was out there when I was a boy. New Englanders built large boats, schooners that could make the far banks and race for home with a fare of lightly salted fish that were quickly spread on the flakes, or drying racks. From there, three sorts of cod went out into the world: merchantable, to Europe, particularly the Catholic countries; middling, to the Canaries and Jamaica; and refuse, for the slaves of the Leeward Islands and Barbados. A ship might sell its fish in Spain, buy furniture in England, and return with high profits. But the great trade was in rum: fish for molasses from the Caribbean sugar plantations; molasses brought home to make rum; rum to markets like Africa, where it brought good gold and slaves for the Caribbean and our South. New Englanders came to be known as the Dutchmen of America. New England trade so pestered the English that one high official of the crown was heard talking about "the most prejudicial plantation in this kingdom." And when the British interrupted that trade by sealing off Nova Scotia and Newfoundland and their West Indian settlements from the flying Dutchmen, the act to New Englanders was an act of war. Adam Smith, the Scottish political economist, quoted John Adams as saying he did not know "why New Englanders should blush to confess that molasses was an essential ingredient in American independence." If molasses, then cod.

The Revolution and what came after ruined Yankee fishermen. Some went off to war and never returned. Ships were sunk, captured, commandeered, or embargoed. But the tradition and potential of the industry were so strong that fishing rights on the northern banks became one of the most explosive issues in the peace negotiations that followed. An onlooker at those talks reported that "the fishery cost more trouble, and satisfaction could not be obtained on that point until Mr. Adams told the British negotiators that [the Americans] would never put their hands to any articles without an express acknowledgment of right to the fishery and tolerable satisfaction upon all other points respecting it." The Treaty of 1783 gave the United States that right, enabling its fishermen to fish on the Newfoundland banks and in other waters of what is now Canada; and to cure fish in the unsettled bays and inlets of parts of the Maritime coast. Rows over those rights unsettled relations between the new country and the old for a hundred years. History's hand is up to the wrist in the current dispute between Canada and the United States over who owns what part of Georges Bank.

It is difficult to say when Georges was first fished, or when or what it was first named. Some say it began as Gorges Bank, in honor of Sir Ferdinando Gorges, one of the most ardent supporters of the colonial fisheries in all England. But cartographers are generally of the opinion that the place made its initial appearance in European history as St. George's Shoal, at the beginning of the seventeenth century. Familiarity gradually secularized the name, and in time laziness claimed the apostrophe. The first Georges men may have been whalers. Nantucketers were after sperm whales around there in the early eighteen hundreds. By the middle of the nineteenth century, boats from Marblehead were visiting the Bank for the summer cod. Handlining was the accepted method. Men lined the rail, their space marked off by vertical pegs called soldiers. Down went the long and heavily weighted "Georges gear," baited with herring or some other easily acquired fish. Up came the cod. Tongues were cut out as markers and toted up at the end of the day. The man with the most was the highliner, then as now the best or luckiest. It might take half an hour to muscle a big cod from a deep bottom in a heavy current, but highliners could and did cut a couple of hundred tongues in a day. At first the ships drifted. The captains were afraid that if they anchored when the tide was running they might be drawn under before they could cut loose.

The Gloucestermen followed after. They accepted the odds and anchored on the Bank. They began taking halibut, the great flatfish, a couple of yards long, and by the 1830s had made the Bank a major fishery. Before the guns flamed at Fort Sumter, Gloucester had taken to icing its catch and sending it by rail to Boston, which was rapidly becoming the marketing center for New England fish, and on to New York. In two decades, the halibut fishery on Georges had been worked almost to death; it still has not recovered. Fishermen turned to the traditional salt cod and, when refrigeration improved enough to gain the trust of the housewife, to fresh fish. Some boats experimented with swordfish, harpooning them in summer as they lazed in from the edge of the Gulf Stream over a hundred miles seaward of Georges. But the stock that benefitted most from cold transport was the haddock. Fishermen considered it too soft in the flesh for proper salting. When icing became common, boats headed out to Georges to take haddock with trawl lines — multihooked lines baited with flanks of salted menhaden and fished from dories. When the otter trawls, the great, grinning, winch-operated nets, came in around 1900, haddock, once trash, became the favored fish. And most of it came from Georges.

Marblehead, the queen of New England fishing ports

in the early days, went into decline, sapped by bad catches, bad markets, bad storms. New Bedford, the place where Melville's Ishmael began a journey that ended atop an empty coffin floating on an empty sea, saw its whaling turn consumptive but hung on. A century ago, the feeling in New Bedford was that the newfangled rock-oil, the stuff just coming on the market that they called petroleum, wasn't all that fine. "The demand for sperm oil and whale oil, as well as for whalebone, will never cease," wrote one optimist, "for there are uses to which these products can be put that cannot be met by other oils or substances."

Marblehead's crown passed to Gloucester, which paid dearly for it. It is hard for a landsman to figure Gloucester, so lovely in the sweep of its hardwoods down to the sea, so profligate of the lives of its citizens. No one knows how many men and boys — fishing vessels usually carried one or two youngsters as apprentices — were lost from the town. One estimate, which seems high, puts the figure at 30,000 since the beginning 350 years ago. Ten thousand seems more like it, and still appalling. A good many went down tumbling in the wildness of Georges Bank.

On the night of February 23, 1862, there were a hundred sail or so, mostly Gloucestermen, fishing Georges. They were at anchor and, as usual, tightly bunched together in shoal water where the cod lay thickest. The weather turned sour, and the wind went around to the northeast. There was no question of hauling in the anchor if things got bad, since that could take more time than wind and wave would allow. Hatchets were broken out to cut the anchor lines, if it came to that. One account (in *"Fifteen Ships on George's Banks"* by L. D. Geller, Cape Cod Publications, 1979) says that by eight o'clock the snow was driving flat across the sea. "Depend on it," a fisherman told his young shipmate, "we're going to have a tough one out of this, and I shouldn't wonder if you had a chance to see more o' Georges than you'll ever want to see ag'in." He almost did. One vessel lost her anchor during the night and drifted within a few yards of them. The next day was worse. Another ship bore down on them. The account proceeds in the high-blown lingo reserved in those days for thumping disasters, particularly of the maritime variety. "A moment more," the young hero reported, "and the signal to cut must be given. With the swiftness of a gull she passed by, so near I could have leaped aboard, just clearing us, and we were saved from that danger, thank God. The hopeless, terror-stricken faces of the crew we saw but a moment, as they went on to certain death. She struck one of the fleet, a short distance astern, and we saw the waters close over both vessels, almost instantly, and as we gazed both disappeared."

Thirteen Gloucester ships went down in collision or lonely foundering. Two more schooners were abandoned. The records are imprecise, but something like 138 fishermen drowned in a night and a day. This less than a year after the *Cape Ann Advertiser* wrote, "Thus far this season [1861] there have been forty men lost who were engaged in the Georges fishery. In view of these losses, we earnestly hope that those engaged in the fishery business will take into consideration the entire abandonment of Georges winter fishing. It has been pursued at altogether too great a risk, and human life is worth altogether too much to be thus sacrificed."

The losses were acceptable — at least, enough fishermen accepted them to keep the fishery operating. They were willing to risk being driven into the welter over the shoals by an easterly gale, drifting to death in a dory, sinking boots first in liquid ice as their ship was rolled on her beam ends by what an early captain on the Bank called "a great knot of the sea." They were willing because of the calling, the challenge, because that is what they had always done, because they felt they could do nothing else. The *Gloucester Telegraph* noted that but for the "unusual losses," 1873 would have been a moderately profitable year. "So far as the fishermen are concerned, whose lives have been spared, the business has yielded good returns." Good meant a few hundred dollars, something above the poverty level, but not all that much.

"So bid adieu to Georges Bank," ends the ballad of the widow-making gale, "Dry up your tearful eye. / Prepare to meet your God above; / and dwell above the sky."

Andrew German, the historian at the maritime museum in Mystic, Connecticut, says that the schooner was the vessel of choice for almost two hundred years, from 1749 to about 1930. The most popular vessel on Georges at the time of the 1862 storm was the clipper, with hollow-ground bows and hulls not noted for depth. It was fast and just right for Gloucester's shallow harbor, but it had the nasty habit of capsizing in steep seas and it was not particularly adept at sailing close to the wind to escape the shoals. Seventy-six were lost on the Bank just between 1850 and 1871.

Gradually, vessels grew larger and deeper and safer. But just as first-rate windjammers began coming off the ways, up popped the seagoing engine. Steam powered some inshore vessels in the 1870s, and some fine, steel-hulled steamers, modelled after North Sea trawlers, worked Georges for a few years around the First World War. But by the 1920s, diesel-powered otter trawlers, named for the otter boards that stretched the nets wide over the bottom, were taking over.

Invention is the mother of necessity. The new hulls cried out for new gear, and it came — some, like the otter trawl, from Europe, some, like the purse seine, from the mackerel fisheries at home. The development of heavy steel dredges made it possible to go beyond the sweet but small bay scallops and take the much larger sea scallops on Georges. The use of transparent monofilament increased the efficiency of the gill net, a curtain of mesh large enough to admit the head of a swimming fish but small enough to catch in the gills when it tried to escape. Big traps rigged with plastic netting opened up the profitable deep-water lobster fishery in the canyons of the continental slopes. And the flatfish — the yellowtails and soles, dabs and blackbacks — swarmed to market when the trawl nets came to the offshore bottoms. Good captains have always been able to find good grounds by reading the sea and looking at the compass. But no amount of experience could beat the new navigation — the electronics in the wheelhouse that can put you within meters of where you want to be in a fog so thick it is better chewed than breathed. No amount of hunch could produce catches like the fish finders that show schools hanging in long blips over the bottom.

A hundred years ago, a fisheries study reported that "many of the fishes and invertebrates which in Europe are highly valued by the poorer classes are never used here. Only about 150 of the 1,500 species known to inhabit the waters of the United States are ordinarily found in the markets." The markets have broadened some since. The redfish, known to the Maine lobster fisherman as an excellent bait, is now doing well in the supermarkets of the Midwest under the name ocean perch. At a time when the cost of fuel and gear has pushed fish prices up to the cost of good beef, consumers are experimenting with less expensive items like pollock and some hakes. But the greater success seems to have come in the methods of presentation rather than the fish presented. Sixty years ago in Gloucester, Clarence Birdseye developed his quick-freezing methods. From his process came the fish stick, a staple in the aisles where Gorton's and Mrs. Paul's — made mostly from imported fish blocks — sit in frosty stacks next to the Weight Watchers.

In 1980, fishermen caught about 294 million pounds of fish and shellfish from Georges Bank, worth a little under $200 million dockside and considerably more when processing and marketing are thrown in. The Americans, fishing all over the Bank, took three-quarters of that poundage and two-thirds of the value. The Canadians, restricted by joint agreement to the eastern third, took almost all the rest, mostly scallops. The American haul from Georges works out to about 8 percent of total U.S. landings and perhaps twice that if the large amount of fish used for industrial purposes, like menhaden fishmeal for chicken feed, is not counted and only food fishes are considered.

Fishermen of the industrial world are its last hunters and gatherers. But their gear is now electronic, and they are getting to be frighteningly good at their business. A while ago the Europeans came back to the western North Atlantic in fleets of ships that would have staggered John Smith and Ferdinando Gorges. The trawlers were as big as coastal freighters, the processors turned catch into product in seconds. The boats worked Georges Bank like so many combines. Things are more under control now, but the technology is still years ahead of the kind of knowledge necessary to manage the resource. Fish are not wheat. They cannot be farmed by the quarter-section. Their abundances and limitations obey imperatives, many of which are not understood and few of which respond well to the kind of harvesting most stocks are subjected to today. Surges occur here and there, but in recent decades, the catch has been coming down.

Things are still acceptable — sometimes, very good. Before the present scallop slump, skippers in that fishery could make very good money — sometimes around $100,000. A reasonably good trawler captain can walk off his icy deck in December with more than $50,000 for the year. You aren't going to hear too many of them talking about their take: someone from the Internal Revenue Service might be around. Besides, it is not their style to be publicly positive. They prefer the salty blues. Things

are no damn good. To go out there, you've got to love
fishing or be good for nothing else. If crap were gold, fishermen would be born without touchholes.

Not many people know more about the fishing life in
New England than an anthropologist named Susan Peterson. She works at the Woods Hole Oceanographic Institution and is married to the chairman of the biology
department there. Susan is in her midthirties but looks
eighteen in the face. She knows the old side-trawlers
with only a bucket for a toilet. She knows the ports and
the men, the lumpers who empty the fish holds at the
docks, the processors, the senior fishermen who run the
boat owners' associations.

Peterson and Leah Smith, an economist, have written
a section for the book the Oceanographic is preparing, a
synthesis of scientific findings and policy issues involving
Georges Bank. The two women report that fuel costs can
represent half or more of the expenses for boats going
far out on Georges. Revenue is up in some parts of the
fishery; boats fishing Georges gross from $100,000 to
$1,000,000. But the cost of new boats and gear is de-
pressing. Most captains own their boats, by themselves or
with their families, but purchase prices are now so high
that financial institutions increasingly have a hand on
the helm. And with the mortgages come interest rates
that can rise to brutal levels.

One result of big debt is that owners burdened by it
tend to be less adaptable when it comes to adjusting to
fluctuations in the fishing stocks; they stick to what they
have always done best, even after returns begin to
shrink. The same reluctance to change can proceed from
the ancient custom of paying the crew by the lay or share
system: so much for the boat owner, so much for the captain, engineer, cook, so much for the deckhands. A year

of good trips can produce $30,000 or more for those who tend nets and gut fish. But lays don't lend themselves to innovation, especially when new methods require trial-and-error periods that may reduce catches temporarily. The fishermen's unions are strong enough in most of the big ports to make owners think twice before taking a flyer.

The fishery, Peterson says, is the sum of hundreds of individual preferences, each captain making his choice on the basis of the boat under him, what his crew can do, what the market may do. Distance counts, too. If you're going all the way to Northeast Peak for scallops, you can figure to be gone for ten days or two weeks. If you're going closer in on the Bank, you can halve the time. If you have the boat and gear to go where you will, and your will is exceptionally strong, you will be away from home and family perhaps three days out of four. Absence, the being away, is more a part of the life than the sun on the sea. Gloucester is still the region's premier port in the number of boats fishing outside the state's three-mile limit and in the number of fish landed. Gloucester has ruined itself more than once by clinging to a single catch, Peterson says, like redfish or whiting. But the lesson has evidently been learned to a degree; there is diversification now. The town is closer to Georges than its closest competitors, and so the trips are somewhat shorter and the fish thus somewhat fresher.

New England's leading port in the value of landed catch is New Bedford (which is third, nationally, behind Kodiak, Alaska, and San Diego, California). Scallops have played a large role in New Bedford's new wealth. Evidence of the old is still around in the narrow streets, some cobbled, in the chandleries and the dark sheds by the waterfront. To Herman Melville, New Bedford was "perhaps the dearest place to live in all New England. . . . Nowhere in America will you find more patrician-like houses; parks and gardens more opulent. Whence came they? . . . Go and gaze upon the emblematic harpoons round yonder lofty mansion, and your question will be answered." To Sue Peterson, the old bones support a new body.

It is no longer a Yankee body. Drownings and dearths and the attractions of inland occupations reduced the numbers of British descendants and left room for late coming. To some degree, all New England fisheries have experienced the influence of outsiders. Smith and Peterson found that among captains of vessels capable of fishing on Georges at some point in the year (which generally means being over fifty feet), fewer than half considered themselves native New Englanders. About a quarter said they were Italian, 13 percent said they were Portuguese, and 12, Norwegian. The rest identified with a scattering of nationalities. Three-quarters of the Gloucester fishing captains — and almost all those found offshore — come from families that immigrated from Italy. Of these, many are or prefer to think of themselves as of Sicilian extraction.

The mix in New Bedford has got to be fascinating to an anthropologist: original inhabitants and Norwegians and Portuguese and a dollop of Latvians. The first Portuguese came in from the Azores as replacement crew on the whalers. Now they make up over half of the people in town, and you can hear their pleasant lisps and nasalities up and down Johnny Cake Hill.

Around 40 percent of New Bedford captains are Yankees, Susan says. "Perhaps because they have been here longer than any of the others, they fish for species considered to be typical of New Bedford — flatfish, mostly yellowtail. Their style of fishing is conservative. They say, 'Squid? Ocean pout? Why catch what you can't sell?' They depend on their families for support, as all fishermen do, but they are not clannish. They sign on anyone who will work for them. You'll find a lot of Portuguese in their crews. And they rely on banks rather

than family financing for their boats; I guess they'd
rather deal with a loan officer than their fathers-in-law."

The Latvians are the scallopers of the place. "When
the local stocks declined in the midsixties," Susan says,
"most of the New Bedford fishermen left scalloping and
went to dragging. The Latvians stayed, often going as far
south as the Carolinas or north to Brown's Bank to get
good catches." They made very good money on the next
upswing of scallop stocks. They have invested it in more
and better boats, and now some of them take their relative ease as captains ashore while others fish their vessels.

"What the Portuguese fishermen have been doing can
best be described as an end run," Susan says. "Instead of
competing with the other groups in the port, they fish for
a variety of species such as squid, or trash fish when it
has been profitable, and assorted inshore and offshore
flounders." They did not sell their catch at the New Bedford auction until relatively recently, preferring to make
their own arrangements with buyers that would give
them a return without the hassle of excessive competition. When it comes to buying a boat, Susan says, "the
Portuguese strategy is to borrow money from as many
friends as possible, set up a company in which they are
all stockholders, and accumulate a down payment. Then
they go to the bank."

The Norwegians believe in equipment. A lot of them
came here in the first half of the century and retain ties
with their families back along the fjords. They prefer
steel-hulled stern trawlers like those in the North Sea,
and they use them to range far out after cod, haddock,
and pollock. Susan says they have learned when to sell at
the big Boston auction and when to land fish at New
Bedford for a better price. Because their boats are designed to be flexible, they can also fish for flatfish. Fathers turn over their boats to sons, and brothers crew for
brothers. The Norwegians are the most highly educated

fishermen in New Bedford; many graduate from college. "They usually spend more days at sea than any other group," says Susan, "and they are the ones most likely to fish on Georges Bank."

It is fiery cold, this January noon, in the little port of Fairhaven across the harbor from New Bedford. The dock by Hathaway's Machine Shop is six inches deep in ice. Floes clog the harbor, and frost smoke hangs over the gray water.

She lies outboard of an old, rusty trawler, and she looks deserted. No, there's a head showing in a window of the pilot house, and it's reddish. They told me over in Leif Jacobsen's office in the settlement house, where they cash the fishermen's checks, to look for a redhead. That would be Lars Sovik, Leif's son-in-law, captain of the ninety-eight-foot stern trawler *Valkyrie*.

Lars comes across the old trawler's deck to give me a hand, smiling slightly at the recording machine and the notebook bag hanging from my neck. We jump down on *Valkyrie*'s deck by a drum bearded with the yellow chafing gear of a tightly reeled net. Two men sit on a bench in the tiny mess. Lars walks by them and down a ladder forward. "Did you bring a sleeping bag?"

"No."

"Well, wait." There is a "v" somewhere in the beginning of the words. Lars returns and gives me a blanket, an extra one of his. "You sleep with the Icelander," he says, and leaves.

I shake hands with the two in the galley: Stanley, the engineer, Norwegian, with the face of a student and a round body; and John, one of the fishermen, a Portuguese, weathered and quiet. Dave the cook is out buying eggs, Stanley says. Oley the mate is around somewhere and the other two fishermen, Tony and the Icelander, will show up any minute. What's the Icelander's name? "Call him Halley," says Stanley. "I can't handle those last names."

I go up the ladder aft of the galley to the wheelhouse to find a place out of the way of the work. There is just enough space forward of the chart table to stand and a couple of handholds to keep me there in a heavy roll. The wheel, with its instrument console, is a few feet away, centered below the wheelhouse windows. The fish finder is to the left, the automatic pilot and a broken sonar to the right. Equipment for two radars is mounted on the overhead, along with two radios. The lorans are over the chart table. Facing aft below a big window giving on the fishing deck are the trawl controls. On the forward bulkhead is the builder's plaque: "Goudy and Stevens, East Boothbay, Maine, 1967." *Valkyrie* is getting on.

Lars comes and sits on the cracked red plastic seat of the conning chair. He winces as he settles as if his back bothers him. His voice, when he uses it, is deep. He says he went to sea with the Norwegian merchant marine when he was fifteen. That was in 1954. After four years of that, he came over here and has been fishing ever since.

Valkyrie can hold eighty tons of fish, Lars tells me, and maybe twenty tons of ice for them. "But we have to stick to the quotas we're allowed." The quotas are worked up by the New England Regional Fishery Management Council, one of eight around the American coast created when the United States went to a 200-mile fisheries jurisdiction back in 1977. "We get twenty thousand pounds of cod, twenty thousand of haddock each week. A couple of thousand more pounds for the state waters." He shows me where we're going: the Leg, a bent finger of shallows out on Georges by what they call the Southeast Part. They got over thirty thousand pounds around there last trip, "fourteen or fifteen thousand the first day. Then come a gale o' wind and when it cleared the fish were gone. Too late in the trip to go further east looking for them. So when we heard another storm was coming the next day, we said to hell with it and went home."

The captains used to call the bottoms by their proper names: Little Georges, Cultivator Shoal, Billy Doyle's Hole, the Leg. ("Old fella got the marine operator to call home," Lars says. "She asked him where he was, and he said, 'I'm in the Outside Hole,' and she damn near disconnected him.") Loran came in after the war, providing an electronic fix from the master and slaves stations ashore. First model A, now C. Now you don't go to Billy Doyle's or the Outside Hole, you go out on a loran line. "We're going out on the twenty-four fifty line," Lars says, poking a finger along the chart. "Maybe twenty-five hundred."

Oley the mate comes in, small and neat, moustache trimmed, smelling of after-shave. "Let's go," Lars says and starts the big Caterpillar diesel. We go astern, past *Elizabeth, Viking*. The high-hulled scalloper *Eagle* is out in the harbor ahead of us. We slide by the *George S. Patton*, also owned by Lars' father-in-law, and *Sippican*, still flying her steadying trysail, and *Poseidon*, and *Marissa*, just in, with a veil of ice on her bows. We follow *Eagle* out to the narrow opening in the seawall. We mutter through and on out into Buzzard's Bay, flat as a puddle. Way off to starboard is a statue on a rock, arms spread like a Christ, like a cormorant drying its wings. It could be the monument to that old sassafras hunter Gosnold they built down at Cuttyhunk, but that's pretty far to show, even on a frozen crystal day like this. Dave the cook comes up for a look. He is a dry and wiry man who used to skipper small boats fishing inshore. The rest of the crew are in their bunks, storing sleep against what is coming. "Hope you have a strong stomach," Dave says to me.

"Well, I'll probably get sick," I say, "but it won't be from your cooking."

"I remember the first trip I made," he says, looking straight ahead. "Wished I hadn't come. About this time of year." He shakes his head and goes back to his galley.

Lars steers with his slippered feet, taking it easy as we saunter through the tidy part of the trip. The full sun warms the wheelhouse. Oley leans against the bulkhead. John the fisherman sits in the sonar chair. He has a cigarette going — everybody smokes heavily — and the sun catches the scarred gold of his wedding ring as he tends his butt. Nothing to do yet. Outward bound.

Lars is talking about the quotas. He is an even-tempered man outwardly, but he heats up a little when he talks about the limits on him. Watches are now eight hours on and four off, he says. They used to run six and six when they could afford to carry nine men before the quotas. Lars winces as a roll gets his back. "The people who want the quotas are the people who can't make them. A lot of us wanted sixty thousand pounds limit, but the council said no, so it's forty thousand pounds a week plus what the state gives us. If you get lucky and go over, you can't bring it in."

We're through Quick's Hole in the fence of the Elizabeth Islands. Lars doesn't use Woods Hole. The insurance people don't like the rocks and current there, he says. Oley has been listening to the radio. "We're in for a spell of weather," he says. As we fight the head tide through the island sounds, Vineyard and then Nantucket, the clutch of boats following us thins out. We find out later that the forecast turned them back.

Lars tells me that *Valkyrie* made thirty-two trips last year, each of five or six days. Before the quotas, he'd stay out about a day longer per trip on average. He likes to fish for the whale cod, the big ones, when they come in early in the year, around the sand heaps in thirty or forty fathoms of water. Not much haddock there, they don't like that terrain. "The whales are in for a month or so," he says, "and then they leave. Nobody knows where they go." He eases his broad, hard body against the chair. "In the spring of the year, they close off a big piece of bottom for the spawning." Then he goes after cod and flatfish in the rips of Nantucket Shoals.

Lars has experimented a bit. Once they rigged *Valkyrie* as a scalloper, but neither he nor Oley liked that. "Get a little breeze of wind, she's wet in the stern. And all the scallop shells flying around." Prices are high, but it's hard and getting harder to find what you're after. Once he tried squid, when the National Marine Fisheries Service asked him to. "They paid us twenty-five cents the pound and cut it from that, and we couldn't make any money that way." So it was back to what they like and what can pay for the rising costs, like the fuel. A seven-day trip to Georges aboard *Valkyrie* usually works out to 4,500 gallons at $1.09 a gallon.

Dave is back. He has heard about the weather. "If only somebody would pay me to stay in till April."

"We can't afford to stay in," Oley says. He grins: "The Coast Guard'll save you."

"They're broke. They can't come out." Dave goes on about running back from Georges under the muzzle of the blizzard of '78.

I ask Oley about how much weather *Valkyrie* can take and still fish. "Oh, maybe fifteen-foot waves, depending on the bottom. If you don't hang up you can fish in a bigger sea. But last trip we hit these big, sharp heaps. You try fishing those, you peel the top right out of the net."

Dave says, "Must be something wrong with anybody who wants to go fishing."

"Well, maybe you better make some pea soup," Oley says. "That'll calm you down."

Supper is ready early. It will be at seven-thirty when we're fishing, to accommodate the watches. Just before eating, we pass Nantucket, the light at Great Point, the town sending up its own pulse of light into the winter. Dave has crowded the long, narrow table with lamb, mashed potatoes, carrots, cake, pie, juice, milk. No pea soup. Everybody eats rapidly. No talk. There rarely is when you eat at sea. Just ten minutes of jaws working and long reaching.

After supper I meet the Icelander in our tiny cubicle. He is in his midtwenties, well built, with a smooth face and a serious smile. At first I think he is speaking with excessive politeness. It is partly that and partly that he speaks his flawless English with excessive care. Halley tells me that he was a journalist in Iceland. And that he got away from two traffic tickets in this country because the state troopers didn't want to bother trying to write down his name. It is Hallgrinur Bjorgolfsson. That is why the crew calls him Halley. "I hope you don't mind that I put my gear with yours in the locker," he says. "I don't want my good clothes to get mixed with the smell of the fishing."

When we are lying in our bunks, Halley tells me of the British-Icelandic cod war. That was polite, too. The Icelandic patrol boats were very solicitous about the British fishermen working in what Halley's people consider their sea. They would radio them, Halley says, and say, "Please go aft. We are going to shoot you some in the bow."

Second day. Not so cold out. Flat sea. The first set is out, and we're trawling at something under four knots. I am back at my stand, wedged in forward of the chart table. From below us come the smells of coffee and frying and "Ya, ya" and "Yes." Up comes Oley in oilskins, laughing. "Cook wants to go home."

Dave is right behind him. He says what he always says at mealtime. "Sumpin to eat?" He takes the wheel while the others have at his table.

We've been trawling a little over an hour, and there is a half-hour to go. One hundred and forty-two feet straight down, way astern of us, a Yankee-41 net made in Germany is fishing, its rollers keeping it almost a foot off the bottom and its floats lifting the headrope, or upper lip, another eight or nine feet. The sand waves are high here.

The vertical exaggeration of the bottom profiler makes saw teeth of them.

Oley is in the conning chair, Stanley in the sonar seat. They have the sea stare, when the mind lets go of the clock and the thoughts roll with the boat. After a long while, Oley asks me about the oil rigs on Georges. This is January of 1982. Four rigs have been drilling since late last summer. Oley hasn't paid much attention to them, he says. They're down to the southwest, where *Valkyrie* rarely fishes. Who's aboard them, Stanley wants to know. Mostly Louisianans, I say, a few Texans. Don't know much about them, Stanley says. All that country music. He twangs a thumb across his paunch. He saw something once on television about that fella Red Adair that makes a lot of money putting oil-well fires out. I tell them the one about how this Easterner at a Houston party begs to meet Adair. He is introduced, obviously star struck, and as he is leaving he says, "By the way, how's Ginger Rogers?" Silence in the wheelhouse.

Oley says he'd hate to see a spill. "But the thing I really don't like is the garbage they leave on the bottom, the pipes, you know. Like when we were starting off in the North Sea. A lot of fishermen hung up on that stuff."

Lars comes up. The bottom is dropping off slightly, and he goes to the trawl controls and slips the cables until he has seventy fathoms out. This is good bottom, he says, and you want enough scope to fish it right. In fifteen minutes, he starts up the trawl motor to haul back the net. Slickers come on below, and seaboots. The winch whines, and the cables come in over the stern. Two fishermen are waiting for the doors by the gallows, the vertical members of the steel gantry that frames the stern. They leap from the sea, these lozenge-shaped, fifteen-hundred-pound paravanes, and they clang hard against the steel. Securing them can be dangerous in bad weather, stopping them off and leading their cables over to the eyes in the winch wires. Tony and Halley work as if someone is counting cadence. In less than thirty seconds, Lars gets the signal to haul back again.

Stanley goes out to the boom winch, and they begin to bring the net aboard, taking a bight with the line from the boom, hauling in a section, taking another bight. A huge skate hangs high in the net, caught by the tail in the mesh. The cod end of the net appears, spouting seawater. The fish are massed there in a knot four or five feet across. Stanley guides the bag over the checker boxes, pens fashioned of movable boards. Halley ducks under and pulls the bucket string pursing the cod end, and the fish slap and slide out of the net to the deck.

Lars spins the wheel to starboard, off on another line. Tony and Halley work over the catch with pickers, short metal rods with a spike at a right angle on the end, throwing cod here, haddock there, yellowtail there. The skates go through a vent in the gunwale back to the sea. There are a lot of them. Oley is out there working now. The men go to cutting. One cuts the throats. Another cuts up the belly and pulls out the viscera. "This cold," Lars says, "you don't have to gill the cod and haddock, just gut them." The flatfish die as they are. "They don't have too much of a gut," says Lars.

The fish end up in the washing basket, a rectangular colander lying in a tub of water. The basket holds seven hundred pounds full. The catch takes up half of it. Pretty small. Stanley winches up one end of the basket, and the clean fish tumble down a small hatch onto a trolley running high in the hold on overhead tracks. Later, a fisherman will finish the work, building layers of fish and ice skillfully in the steel and aluminum pens of the hold. Now, Stanley pulls the net back off the drum with the boom winch, and the men let enough of it stream in the sea to pull the rest into the wake. The idlers come off the drum and are hooked up. The doors drop like knives, and we are fishing again. The fishermen shuck their oilskins, store them in the big locker aft of the galley. They have spent hugely of their energy and wasted none of it and they are hungry.

The second haul is smaller than the first — mostly skate and ocean pout, thick-lipped, thick-bodied eelish creatures. Cod tend to go off the bottom during the day, hanging where only the huge midwater trawls can get them.

Lars heads northeast. In the middle of the afternoon, he finds what he's looking for. He turns the helm over to Oley and goes below for a couple of hours' sleep. The depth finder shows fish just off the bottom. They look like clumps of saplings. We trawl. Oley thinks we may be looking at whiting or baitfish, you never know. He hauls back. "Don't see no bag comin' up. Vait. A little sumpin, I think." The cod end, the bag, breaches like a sea serpent. Some big cod, thirty pounders. The seabirds are like bats from a cave. They fall in squawking clouds, they pile up on the water. Oley is going around for another pass. "Fish can spook," he says, "but if you're alone you can trawl on 'em for a while." He keeps his eye on the fish finder. "They're some spotty."

I try to catch Oley off guard and ask him now, in the middle of the hunt, what he likes best about what he does. It doesn't work. "Oh, fishing a good bottom." A puzzled pause, as if he doesn't think much about such things and doesn't like to. Then, "The best is when the man tells you you've got a trip in the hold and you can hang the doors and go home."

Lars takes over the helm a few minutes before four. We haul back a half-hour later. The corpse of a gull comes in on the wire. Stanley tries to strip him off, but he passes onto the drum to be crushed. There is a tear in the belly of the net. And two tons of fish — a good many small cod — in the bag. They are sorted and cut and lie in the bloody water of the washing basket, eyes bulging at the darkening sky.

Stanley and Oley sharpen their folding knives and tend

net. They feel for the holes in the mesh, pulling and
twisting until they sense what must be cut away. Then
they rebuild, using needles — bobbins, really — to knot
in the new body, trim and tight. Meantime, the others set
out the spare net, and in the starkness of the working
lights the birds show white as doves.

Lars says that a lot of things besides digging into sand
heaps will tear nets. Rocks ploughed up by the scallopers.
Junk thrown overboard. Treelike sponges. Black mussels.
"They're worse than rocks," says Lars. I ask why he
doesn't get the sonar repaired to help him on bad bottoms. "Well, the salesman that sold it left town," he says.
He has forgotten how to use it.

They are still crocheting away on the torn net when
the spare is hauled back. The doors fly toward their stanchions like rusty mantas. A big bag, mostly skates. Lars
says he thinks skates scare fish away, and he goes over to
the chart for new ideas. He points out the areas closed in
the spawning season. "Doesn't give you much room to
work." He runs a finger across the chart to show me the
Canadian claim line. "They want half of Georges. If they
ever get that, we may as well forget it."

The next set catches a lot of pollock. They don't bring
much at auction, but consumers are showing more interest in them now that they are more interested in fish bar-
gains. Lars says pollock are best taken "coming dark or
coming daylight."

The cod go back to bottom during the night. I can hear
Valkyrie setting out every couple of hours, and the trawl
power system makes sounds that bring me dreams of
steam locomotives in the Berkshires, where I grew up.

Third day. A storm came up from the Carolinas last
night. Ten-foot seas and a cold haze. We're snugged
down. Tony says every time Dave makes pea soup we get
a gale.

"If you don't like it out here, go home," Dave tells him.

By noon, there's a following sea and a warmish fog.

The wind is southerly now, but the forecast says it will back around to the northwest and gust over forty-five knots. Oley says Lars told him to "set up to nordard" and fish tonight. Not much point wasting time in this weather daytime fishing.

The radio is quiet. There can't be many boats out here with us. They know there is a big blow coming. The sea is orogenic today. Combers lift the stern high. The day drags, and I remember the long, dull times at sea in the past when there was nothing, absolutely nothing, new for the observer to observe. Don't come to sea to watch. Mark Twain wrote that "a long sea voyage not only brings out all the mean traits one had and exaggerates them but raises up others which he never suspected and even creates new ones." Even Conrad was bored at sea. Even oceanographers. One I heard of at the Woods Hole Oceanographic went 'round the bend after a month of bobbing around on some uninteresting piece of water. He jumped into the sea and, after they had fished him out, said he thought he had seen his wife paddling by in a blue canoe.

Damn the roll. It carries me directly over a gull paddling in circles, waiting for his share of innards. His head, directly below me, looks ridiculous, like the pommel of a misplaced cane. How do they keep their feet warm? *Valkyrie* is bucking so hard the fish finder isn't working properly. The wind is streaking the waves, and it's shifting. "West-southwest," Oley says. "This is what he was waiting for. We don't usually hang around for long. He was waiting for this to blow through."

Lars comes up. "Every time you get an easterly, then a northerly with the glass falling, you've got to have a northwester for the glass to get back up."

Fourth day. Winds around forty-five knots, ten-to-fifteen-foot waves. The radio says it will ease. We are drifting in the troughs of a very serious sea — black and white with a touch of mouldering green. The wind peels the skin off the rollers, thrums in the booms and gantries, blows harmonics.

Dave: "I hate this time of year."

Tony: "I've seen these storms last three days easy. Can't go home or nothing." Running for home in a blow like this, running over the shallows, is not smart.

Lars: "I think we ought to go into the oil business. This is too hard."

We drift and hang on and curse and talk a little. About money. Tony says he made $45,600 last year with two months off. Halley made $26,700 for half a year. About the fish. Lars shows me the catch report he keeps for the National Marine Fisheries Service. He says he believes that cod load up on stones before a storm. Wonder what Bigelow would say to that. About boats and gear. Lars went all the way to Seattle for Fish Expo, the big industry show where they display the latest electronic cod catchers. *Valkyrie?* She cost a half million new, and it would take a lot more than a million to build her now. Unions? The boat has always been union. That means pulling 4 percent right off the top for the welfare and pension funds. Lars would like to get away from the union's power, but that would mean getting out of town.

By six in the evening, we're setting out, still rolling and taking seas in a frighteningly random way. John and Stanley are out on deck readying the net, when the engine dies. We all freeze. Stanley disappears, and a minute afterward the blessed muttering resumes. Oil pressure problem, Stanley says when he comes up from the engine room. "You never know."

You don't. Just before supper, the Coast Guard puts out a call on the radio. There is a man overboard, way north of us. Nobody thinks he could last more than a few minutes in this water. Lars sits a long time in the conning chair, quiet. When Dave comes up with "Sumpin to eat?," Lars gets up slowly. "I've got twenty-four winters on Georges Bank. That should be enough, really."

In the night, we can hear the Canadians up by Northeast Peak. They are talking about a huge trawler one of them saw on the horizon, probably one of the few foreigners still allowed to fish under the new permits. He can't fish the disputed area up where the Canadians are, so he must be heading down the Bank to the American federal sea. You should have seen the Spanish pair trawlers working out here before the 200-mile law went through, Oley says. They weren't all that big, but with two boats working that midwater trawl, you ought to see what they hauled back.

Fifth day. Sun and a calming sea. There is a big haul in the checkers, almost five thousand pounds, but the net is badly torn. The needles and knives are flashing. An art, I say to Lars. "They never knew anything else," he says. We go down to breakfast, leaving Dave at the wheel. "We never used to tear up around here," he says in the empty wheelhouse. "Must be them scallopers has been through dredgin' rocks."

The spare net is over, and we're towing against the current. It's tricky. If you give it too much throttle, you'll lift the net right off the bottom. Other boats are out now that the weather is breaking up a bit, and Lars is working to fill the quotas so he can get to auction in New Bedford before they do. Right now, he says, the New Bedford prices are bullshit. Haddock going for over $2 a pound at the end of the week, cod for $.90 to $1.75, yellowtail for $1.25. As against about $.30 when the market has what it wants.

Lars is on the radio, talking to another Norwegian captain, Gabe. Their English escapes me, but I think Lars is talking down his catch. "Hasn't been very good up here," he says. "Two tear-ups. Too rough for you."

"Ya," says Gabe. "One storm after another. You seen Red's boat?"

"Maybe he put in at Nantucket."

"Ya, maybe. I hear some Portuguese came out on the twenty-seven-hundred line after yellowtail."

"There's a sign of haddock up here."

"These fish don't want to stay on the bottom."

"Well, see you later."

"Ya, finest kind."

Valkyrie hauls back. The rag markers come in, ten fathoms apart. Three rags together tell Lars the doors are close in, and he eases off on the winch throttle. We are still taking seas. The fishermen are too busy to look for the big waves. They hear them and make their escape just before it is too late. No sign of life vests. Their bulk could be dangerous for a man working close to the cables and drums and doors. The boat does carry survival suits — toe-to-head cocoons of insulated plastic that can preserve life for hours or days in the water. Lars jokes that he told the salesman to use Stanley as a model but that they couldn't find a suit to fit him. Lars says he has forgotten how to get into the damn things.

Nothing this haul. It's a water tow. Lars spins the wheel and goes over the charts. While he's rummaging around, he finds one that shows the fjords where he grew up. He shows me the pinpoint of a small town at the head of one of them: his. He shows me the city of Stavanger, a booming supply base for the oil rigs and production platforms in the North Sea. "The prices," he says. "Everything is out of sight there."

"Hear the weather?" says some skipper on the radio.

"No," says another, "and I don't aim to till it's time to go in."

"Damn sea broke my window last night."

Sixth day. Bitter cold with heavy seas and another storm coming. *Valkyrie* is covered with rime. The radars are icing up, and the one Lars prefers for close-in work has blown a fuse. Lars is groping in the drawers for a one-amp replacement. He can't find one, so Stanley takes

a five-amp fuse from the big radio. Lars is jumpy. The
closer you get to land, the colder it gets and the more you
ice up. If he ices too badly, he'll miss the Monday auction, the one he wants. He hears me talking to Oley about
oil spills and turns testily: "Don't worry about the oil and
about the fish. Mother Nature will take care of that in her
own way." He looks out at the smoking swells and puts in
the fuse. It works.

Seventh day. Colder still. Ice crystals in the water act
as oil does on the waves. Lars totals up the catch: 5,000
pounds of haddock, 37,000 of cod, 500 of yellowtail, 2,500
of lemon sole, 4,500 of pollock. "We're lucky to get that,"
he says. "But it's not what you get, it's what you get for
it."

There were moments last night I will not forget. *Valkyrie*'s roll got more pronounced as she iced, and she
hung, heavy, before swinging over. Dave got so nervous
he forgot to bitch. Lars was slamming the seas, trying to
make that market just sixty miles away. The ice won. He
had to throttle back to keep the spray from freezing so
thick on the bows that it went past the danger point. We
had ice mallets, but it would not have been pleasant to
chip. Lars and Oley kept the radar working by dousing
the base of the revolving arm with hot water. Now we're
making ten knots in Nantucket Sound in a flat-ass calm.
We're so iced up we look as if we belong on a cake.

Lars calls home and finds out that the Coast Guard ice-
breaker is working in the harbor. *Valkyrie*'s steel hull
will have no trouble, but the old wooden-hulled trawlers
will have to watch themselves this frozen day. We come
through Quick's Hole again into light pancake ice in
Buzzard's Bay. A big Coast Guard cutter ploughs
through it outside New Bedford, leaving a black wake of
open water. "I've never seen cold like this," Lars says.

Eighth day. New Bedford holds its auctions in a handsome, small brick building on the town pier — scallops at

seven, groundfish at eight. *Valkyrie* is the only boat on the board, so there's next to no auction. Tichon, the big processor in town, buys the catch. Halley is in a room separated from the auction room by big glass windows. Stanley is there and John. Lars comes in later. About twenty men in all are standing around. A few of them are talking about the fisherman lost overboard. Halley tells a friend, "You got to accept the possibility that something will happen to you. If you die, you die." "Yeah," says the friend. "I used to think that before I went to Vietnam. No more."

Halley and I drive over to the processor's dock. Lumpers, the longshoremen of the fishing industry, are on board, and canvas buckets are swinging down to empty the hold. Halley and Dave take turns standing by the chute where the buckets are unloaded and the fish packed into boxes. They count the boxes and make sure the lumpers don't put more than a hundred pounds in each.

The Fisheries enforcement officer goes aboard to check the catch. After he leaves, Stanley comes over. "Too much regulation," he says. "That's why the fishing fleet is in such serious trouble. You catch some species, you gotta throw them overboard. You bring them in, you pay a big fine." He looks at the three inches of ice on the wheelhouse. "Hope that melts before Friday." Friday is when they are going out again, in three days.

Halley gives a lumper a hard time for throwing a big cod into a box already almost full. "What's your trouble?" says the kid. "You guys get these for free."

When the catch is ashore, Halley rides with me back to downtown New Bedford where he shares an apartment. Tichon always buys *Valkyrie*, he says. "We've got a reputation for the best fish. We're finicky about how we pack it. And we really did get some whales this trip. I was surprised. It was so early." He is angry at the lumpers for the

Oil and Water

way they mishandled the fish. "Back home, we never pick them in the fillet, always in the head. Here, nobody cares."

Halley asks me, journalist to journalist, what it was like when the first oil rigs came to Georges. As I tell him about that, one memory keeps twinkling away behind what I'm saying. I was out on Georges aboard a service boat when the second rig showed, under contract to Mobil, towed by two seagoing tugs. It looked like Eiffel afloat. It was fresh from the Gulf of Mexico with a transit crew of very southern individuals.

A Yankee voice came on the radio, indignant, a lobsterman: "Wheah ah yew anchorin'?"

They got that. Back came the coordinates.

Back came the Yankee: "Yew ah well away from my geah." And, inevitably, "Finest kind."

Nothing from the rig. Nothing from the tugs. Then the newcomers to those cold seas spoke among themselves.

"Wuddeesigh?"

"Dunno. Wuddeeyoo thaink ee sighyud?"

"Ah thaink ee sighyud pah inna skah."

3

See You in Court

THIS one nation, indivisible, was divided once, ripped
apart over what one side regarded as an essential resource. We've grown since into a continental country
with bizarre extremes of geography and ethnicities past
the mention, and though we remain surely one of the
world's most pugnacious societies and among its most
violent, we no longer seem to harbor the kind of divisive-
ness that turned Bull Run rusty with blood. Canada has
more problems than we with separatism, and so do Spain,
Belgium, India, many others. But if the flags of secession
ever do flutter again over this one nation, indivisible, the
chances are that some claim to some precious resource
will be among the casus belli.

Regional resource disputes are increasing in America,
particularly those in which access to energy is involved.
Many are western, like the one over the tax Montana has
placed on its coal. Georges Bank is special because it is
eastern and, far more important, because it involves *two*
primal resources, one renewable perennially, the other
renewable only over geologic time. The fight over who
does what out on those swirling shoals is in part a fight
between those who fill the nation's fuel tank and those
who fill the nation's belly, in part a fight between a na-

tional good (increased energy supplies) and a regional
bad (threats to the fisheries of New England and to its
beaches, so richly sprinkled with tourists). Now that national growth no longer looks as desirable as it once did,
now that we are falling in on ourselves trying to reslice
the old national pie, battles like the one for Georges, with
all its hang-in-there, keep-a-goin', resolute defiance of
any resolution, could become far less special, far more
common.

Paul E. Tsongas, a liberal Democrat from Massachusetts respected by many of his more conservative colleagues in the Senate, wrote a foreword to a book called
Regional Conflict and National Policy (Resources for
the Future, Washington, 1982). "I think," he said, "that
a major generator of regional problems is that oil is a finite resource. That is a reality around which everything
else revolves. Politically, the Republicans would argue
production, production, production — that the way out
of the dilemma of a finite diminishing resource is to produce it faster. . . . The Democrats, on the other hand, will
tell you that the energy crisis is a function of the oil companies — that somehow if you could put Mary Poppins at
the head of Exxon the problems would go away."
Tsongas thinks both approaches "equally absurd, equally
rhetorical and equally successful" — particularly when
talking to the converted. "And that is basically how most
people address the issue," he wrote. "We are awash in
rhetoric, not to mention hypocrisy, when what we need is
a careful sorting and weighing of the facts and values involved in making — or not making — a decision."

Tsongas was honest about the terrible traps set for the
politician who walks into a resource feud. He had, he
said, chosen the national interest over the regional in
voting for the Alaska lands bill setting forth extensive federal holdings in that wilderness. "Had I been a senator
from Alaska," he said, "I might well have voted the other
way." Still, Tsongas thinks, there are national interests

and national interests. In Alaska, it was conservation. "In the case of Georges Bank, the purported national interest might well be destructive." That is Tsongas' view, and it must be said that he has spent a lot of time running hearings and reading reports on what oil drilling might or might not do to Georges Bank. There are, as will be seen, plenty of other views.

Under the Constitution, the Congress has the task of overseeing the management of our public resources. It has done this by delegating management responsibility to agencies like the Department of the Interior, whose authority over oil and gas, timber, coal, and other desirables on and under public lands (amounting to about a third of the total American acreage) makes it one of the most powerful political clubs in Washington. Another manager, far weaker, is the Department of Commerce, which regulates commercial fishing. In early years, the agencies wobbled from innocent mistake to massive scandal. Since the middle of this century, though, they have become much more sophisticated and, they say, much more efficient at the game.

The Department of the Interior has also become amphibious, following the oil men out onto the continental shelves. These are the rims where cod and haddock play, where men ply the richest fisheries. They are also, in spots, immensely attractive prospects for the drill. Oil technology can now "make hole," as drilling is often called on the rigs, and recover hydrocarbons — at a profit — from bottoms over one thousand feet below the surface; in a little while, the limit will be considerably below that. The Gulf of Mexico is still producing from its thousands of wells, though it is an old play. The North Sea has dramatically changed the revenues of Norway and Great Britain, though it too is passing middle age. Oil is finite. New finds are needed, or will be when the current oil glut has passed. There is considerable interest in formations off Africa and South America; very

special interest now in what might lie off the People's
Republic of China; and a range of interest, depending on
the oil company, in Alaska, California, some unprobed
parts of the Gulf of Mexico — and Georges Bank. It is up
to Interior to translate that domestic interest into lease
sales on the outer continental shelf, and to do it in ac-
cordance with a body of law and regulation and stipulation that can and does give the most seasoned bureaucrat
or jurist a case of the dwindles.

Oil men could pretty well call their shots when the first
well in seawater went down off a wooden pier in Sum-
merland, California, and became quite the thing to see
for the Gay Nineties tourists. The first freestanding drill
device went into action close inshore in the Gulf of Mexico in 1938, but even by the end of the Second World
War, there were fewer than a dozen holes off Texas and
Louisiana. They were enough. Government caught the
scent of revenue and moved to assert its influence before
sand heaps turned to pin cushions. During the war,
Franklin Roosevelt had directed his lawyers to find a
basis for a strong federal claim to the shelves, and Harry
Truman declared that what lay on and under them was
subject to the exclusive jurisdiction and control of the
United States.

A kind of wet states-rights movement developed in reaction. Governors searched their archives for old maritime claims, but only one or two were successful in the
litigation that followed. The Supreme Court gave Texas,
which had been a republic before entering the union,
three marine leagues, or ten and a half miles, of sea. But
the state of Maine, which more recently claimed the right
under old colonial grants to sell oil leases far out from its
shores, was reined right back to the conventional three-
mile limit. Today, with the federal government less willing and less able to ease regional and local stagnation,
coastal states are even more interested in the resources
now beyond their taking. Much of the interest is in pre-

venting environmental damage to marine resources inside state waters or on an offshore bank frequented by
state fishermen, but not a few governors still dream of
what sharing the take from some future frontier bonanza
off their beaches might do to liven things up at the state
house. They look north to the big and potentially bonan-
zoid Hibernia find off Newfoundland, struck by Mobil Oil
Corporation's Canadian affiliate and other operators. Hibernia might be worth seventy billion dollars, the New-
fies say, and they have fought Ottawa for it. So far,
they've had no more luck with their claim than their
western cousins in Maine have had, but they keep on
dreaming.

The federal offshore leasing program got under way
only months after Congress passed the Outer Continental Shelf Lands Act in 1953 — "outer continental" referring to everything from the state limits to a muzzily
defined line just about where the widest shelves pitch off
into deeper water. The act set the rules for the game:
competitive bidding by American firms, the winner to
walk away with a lease that would let him gamble for a
while on a piece of the public's submarine land. Where
conservation was mentioned at all, it usually referred to
measures to minimize damage to the oil resource; the
rules said little about fish.

No one outside the oil patch paid much attention to
the process. It was a more peaceable poker game then.
"The oil companies came to us," one veteran of the early
days of leasing at Interior told me, "and they said they
were ready; we leased on demand." The department's
realtor, the Bureau of Land Management, arranged for a
sale every year or two. In fifteen years, it offered only
about 1 percent of the American continental shelf. And
then came conservation of an entirely different sort.

It would be well within the limits of fictive license to
say that the American environmental movement started
at a particular time and place west of the California

mainland, several miles west, in the Santa Barbara
Channel up the coast from Los Angeles. There, in 1969,
an oil well lost control and blew, and quite a lot of product
washed up on beaches frequented by some people of
clout and fiber. They formed GOO — Get Oil Out — and
electronic pictures of oiled birds began appearing in living rooms across the country. That was at least a year before inlanders began burying new automobiles to protest
what they thought industry was doing to the environment, and there has been a salty fringe to the environmental movement ever since.

Basic environmental legislation, like the National Environmental Policy Act, clearly had marine implications.
It declared that no government agency could mess with
Mother Nature without first devoting a lot of time and
energy and paper to the preparation of an impact statement that established in full public view the environmental consequences of its proposed action; clearly,
offshore oil operations qualified as messing. The Environmental Protection Agency, itself a child of the earth
movement, felt constrained by the Clean Water Act of
1970 to work up permits regulating the discharge of various chemicals from offshore rigs and platforms. Congress
passed legislation to protect marine mammals and endangered species and to preserve rare habitats through
the creation of marine sanctuaries. It encouraged the
states to manage their marine holdings better by offering
them financial help to do so — if they would develop
coastal zone management plans acceptable to the Secretary of Commerce. (Remember? The United States Department of Commerce regulates our fisheries, through
the National Oceanic and Atmospheric Administration,
NOAA, parent of the National Marine Fisheries Service.)

Then came the Yom Kippur War in 1973 and the en-
suing Arab boycott. Oil prices doubled and quadrupled
and no ceiling seemed in sight. The public perceived a
shortage, and messing with Mother Nature to alleviate it
suddenly took on respectability. Bumper stickers read

differently. "Save the Whales" was still around, but in
company with "Out of Work? Hungry? Eat an Environmentalist." Richard Nixon told the Department of the Interior to offer ten million acres of the outer continental
shelf in 1975 alone — about as much as had been put on
the block over the prior twenty years. Project Independence, the President said, would give us all the oil and
gas we needed by 1980. Gerald Ford continued Nixon's
policy, calling for a five-year leasing plan that would include acreage in the frontier areas of the northern Atlantic and Pacific, including Georges Bank.

That brought New England awake. Some hardy souls
spoke of what petroleum from Georges might do for the
region, which must import just about all the fossil fuel it
uses to keep winter at bay and industry on line. Others
saw little benefit. Since New England had long since
turned its back on major regional refineries, they said,
any oil found offshore would be processed, and probably
consumed, elsewhere. A gas discovery could boost what
was in the regional pipeline system, and that might help.
Oil operations could mean jobs in some coastal communities, but judging from what had happened in Scotland
when the North Sea began to produce, employment
would peak and then begin to turn into a boom-bust spike
that would ultimately bring more dislocation than pros-
perity. These pros and cons should have been given more
of a public airing, but they were overridden by the dra-
matic arguing of the environmentalists: drilling on
Georges might well put an end to the great fishery. It too
was an economic argument, something not often pre-
viously stressed by the environmental side, but an economic argument pleasantly scented with the romance of
the great fishing banks and the call of the boats. It was
also states rights, with that new left hook: in Massachusetts as in California, liberal environmentalists were acting like the old conservative Dixiecrats. Damn feds, they
said, goddam, meddling feds.

Though Gerald Ford had identified Georges as a lease

sale area, it fell to Jimmy Carter to make the moves. Or rather, to Carter's Secretary of the Interior, Cecil Andrus. Andrus is in his middle fifties, a former governor of Idaho. He is tall and holds himself so he looks taller. His face is on the edge of handsome — his nose almost regal, his eyes almost arrogantly steady. When he came in, early in 1977, he let it be known that he was going to throw a loop around the "rape, ruin, and run" resource development operation he thought Interior had permitted under past presidents. To an unusual degree, given his job as the nation's landlord, he paid attention to the body of environmental law as it built up in the federal books.

Andrus has a good deal of the reformer in him. I saw it one morning when I met him in New York after Ronald Reagan had shown the Democrats the revolving door. He sat in his hotel room beside the unmade bed, in his shirt-sleeves and yet with that immaculate look of the powerful. He was talking about his efforts to change how things were done in the offshore. Take Baltimore Canyon, he said. That was one of the first sales on the East Coast, and the oil companies went at it as if they had warmed up in the casinos of nearby Atlantic City. "A billion two hundred million dollars of bonus bids, and I've got a little bottle in my office about that high that got almost all the oil they found out there." Not quite: they found oil, but not enough to produce commercially. "It's not fair to the oil companies," he said, "to have a Baltimore Canyon. Nor is it fair to the taxpayer to give all the hydrocarbons in the Gulf of Mexico to the oil companies for one-eighth royalty" — roughly what the federal government receives on production there. Andrus wanted the industry to explore offshore on its own, as it has always done in this country. If it found oil, "you'd agree upon a percentage of government ownership."

Andrus' phone rang. He took about twenty seconds to

deal with the caller and then dropped back into his groove. "I think that government ownership has to be in excess of fifty percent, if the companies want to deduct all expenses so it's pure profit. But that way, the oil itself finances the whole show. If oil can't be found in commercial quantities, then the federal government shouldn't be entitled to millions of dollars from the oil companies. But if it's a bonanza, the public ought to get its ownership rights." A small smile. "The oil companies didn't like it."

Andrus felt no qualms about drilling on Georges Bank. He told me in New York that there was plenty of environmental protection, but that environmentalists in New England disagreed, some of them vehemently. They used the most ephemeral of government estimates (showing that under a certain set of assumptions there might be the equivalent of one week's national oil supply under the great thumb) as if they were hard data; environmental rhetoricians asked: Is this fabulous fishery worth a paltry week? "New England could sure use that oil in a cold winter," Andrus said to me. "Besides, there's legitimacy to the argument that if hydrocarbons are there, and they can be removed safely, then you've gotta participate."

A friend of mine, an attorney, told me once that in law school he was told that the best law is no law. I didn't know him well enough then to ask if that applied to lawyers as well. From what Cecil Andrus said to me, I would bet that he thinks it might. "You have those people in some of the foundations, nonprofit organizations," he said, riling up. "They often don't have to fund their organizations themselves, and they get to play lawyer. And they'll bring frivolous and unneeded lawsuits simply because they don't have anything else to do." I don't know why I didn't ask the former secretary if, among others, he had in mind the man who has been one of Interior's most artful opponents in the battle for Georges Bank, Douglas Irving Foy, head of the Conservation Law Foundation of New England.

➤ Doug Foy, in 1981, is in quite a prime. He is around six feet, blond, and his longish face is usually set in what you might call classic rugged. He has the appearance of a celebrity; he has the chemistry of a Marlboro Man. But he also has a tinge of Dudley Doright, the virtuous Canadian Mountie in the old "Bullwinkle" television cartoons. Dudley with dignity. Doug doesn't make a practice of falling on his prat the way Dudley used to. He practices climbing rock faces — and takes his little daughter, Emily Bronwen, for scrambles on the easier slopes. He admits to a mania for bicycle racing. He and his British-born wife, Leonie, sit balanced in canoes shooting rapids not designed for that purpose. Doug used to be a competitive oarsman, a fact I found out not from him but from a colleague. He was on the Olympic team at Mexico City. Couldn't I tell? asked the colleague. "I mean, look at the guy. Magnificent!"

A competitor's competitor, but with that sense of delight in the fixed star. In another age, Foy might have been a grail hunter. When he was young, he followed his own ideas about the order of things, natural things, the things his grandmother had taught him on their walks in the New Jersey countryside. When he saw some of his friends throwing stones at ducks, he did what he could to deck them. His language is clear and persuasive and, when he talks about the strategies of his job, full of fists. "Stick it in their ear," he says. "Go blasting in."

Doug thought he would be a physicist. After college, he went to Cambridge University on a fellowship to study geophysics. "But by my senior year in Princeton," he says, "I realized I was much more interested in oral advocacy and argument." So he went from Cambridge to Cambridge, to the Harvard Law School. Public-interest law attracted him, and he worked in civil rights and legal aid. Then the executive directorship of the Conservation Law Foundation opened up. Foy, still, in his heart, the defender of ducks, jumped for it.

The Conservation Law Foundation claims to be the only organization of its kind in the country. Its dozen or so lawyers and scientists and interns work with others of their kind around the region, with politicians and planners, anyone whose expertise can help them use the law to improve resource management in New England. Its funding is mostly local, and when Doug got there in the midseventies there wasn't a great deal of it. "The pace was slow and sleepy, and we didn't do much with advocacy," Doug says. Now the budget has just about quadrupled — to a half-million dollars — and the pace is faster than the Allagash River when the ice goes out. The chairmen are likely to be respected politicians, often Republican, like former governor Frank Sargent, and former state senator Frank Hatch. The vice-chairman is John Teal, perhaps the most politically engaged of the senior scientists at the Woods Hole Oceanographic Institution, a ninety-minute drive from Boston down on the lower shoulder of Cape Cod. The rest of the board is a high-energy mix of old money and new activism.

New England is right for the Conservation Law Foundation, just as it was right for Rachel Carson and *Silent Spring*. It was first to industrialize, first to flourish as the river mills turned their profits, first to feel the damage as the sawdust in the tailraces spooked the salmon, as the coal smoke choked the factory towns and, now, as rain approaching vinegar in acidity moves in from the west to kill conifers along the Presidential Range in New Hampshire. Scratch a Brahmin and you'll find a conservationist, a fundamentalist in the religion of clean air and good water. No wonder the Massachusetts Audubon Society was fighting pollution years before the national body could take its eyes off the groove-billed ani and the copper-tailed trogon long enough to follow suit.

Doug's place of business and battle is in an old brick row house on Joy Street, close by the State House and the Boston Common. The building houses a warren of like-minded groups: the Land Trust Exchange, the Environmental Lobby of Massachusetts, the local branch of the Sierra Club, the Friends of Harbor Islands, the local Friends of the Earth, the New England River Center. All that is missing is a wrought-iron sign over the doorway: "Abandon all hope, ye exploiters who enter here."

Even with all those allies, Foy isn't completely comfortable at 3 Joy Street. "There is a certain irony about being an environmental lawyer," he says. "I spend most of my time in libraries and courtrooms, while what I really care about is out there," tilting his head toward the window. He must be indoors to defend what he loves most, the outdoors, and he has built his private life around the irony. He lives in a town west of the city and bicycles the twenty miles each way except when weather makes the trip suicidal. On warmer weekends, Doug says, "we end up trying to disappear in the mountains."

Under Foy, the Conservation Law Foundation has negotiated and litigated, mostly the latter, in disputes involving off-road vehicles that are scalping the dunes on Cape Cod, a proposed refinery on the Maine coast, toxic dumps here and there, and endangered farmland. But the battle that has put 3 Joy Street on the national map has been the battle over Georges Bank.

Ladies and gentlemen, welcome to the big water fight, New England's amphibious extravaganza, featuring fearless advocates of every persuasion, governments locked in battle, a supporting cast of hundreds: heroic

fishermen facing howling Atlantic storms to bring us the riches of the sea; dauntless drillers from the Deep South willing to risk their lives in those same tempests to keep America warm and rich. Oh, it promises to be a rouser, ladies and gentlemen, splendid.

In that corner of the ring, please welcome the rugged Mr. Foy, the challenger. Thank you. And in this corner, your applause please for the battling hurricane of old, the gentleman wearing the western look required of all self-respecting Secretaries of the Interior, the Honorable Cecil Andrus. Again, thank you. In a minute, the beginning of round one.

Joy Street is a pleasant stroll from the John W. McCormack Building, where Francis X. Bellotti, the Attorney General of Massachusetts, has his offices. Bellotti is an exceptionally independent-minded man, given to following the legal line of an issue rather than the political arguments of whatever governor he may be serving under. He is also a powerful man: he is empowered to take action in court when he thinks the legal rights of the commonwealth are at risk. When Douglas Foy and Francis Bellotti are both interested in a case, their staffs are apt to be in touch often.

In the midseventies, both men were watching efforts in Washington to amend the Outer Continental Shelf Lands Act, the enabling legislation for offshore drilling, then creaky with age. Hearings were held during the Ford administration, and Massachusetts and others among the twenty-four coastal states, along with the oil industry, environmentalists, and fishermen's groups, held forth on what needed fixing.

The states, Massachusetts a clarion among them, protested that they didn't have enough of a say in decisions that might result in significant changes. Ashore, they were worried about sudden economic stresses in towns near oil company supply bases, say, or overcrowding in fishing ports when trawlers competed for dock space with supply boats servicing the rigs. Offshore, the specter for the states was the oil spill. Some harbored an almost visceral fear of what a big one might do, for example, to the Georges fisheries and to tourism, each bringing hundreds of millions of dollars to Massachusetts alone every year. The states and their fishermen wanted compensation for damages inflicted by offshore operations. Some governors also wanted funding so that they could plan for the onshore economic and social impact of those operations.

Environmentalists wanted language in the amendments that would effectively close down a rig or a producing platform if clear and severe damage to sea life were to develop. If that were to happen, the oil companies argued, then the operators who were shut down ought to be compensated. That would help, they said, and so would overall simplification and clarification of the regulatory snarl they now had to deal with.

The urge to amend offshore legislation stalled under Gerald Ford. But when Jimmy Carter and Cecil Andrus went to Washington early in 1977, they smiled upon the amendments. The reformer in Cecil Andrus was delighted with the chance for change, and he worked hard with the Congress to push the changes through. But Andrus was also determined to be fair in his dealings with the oil companies. He knew what happened when the schedules for lease sales slipped for one reason or another — the losses that were inflicted on companies that had borrowed heavily to get their bid money and then had to sit and watch the clock run on their interest payments. He promised the offshore operators that he too would watch their clock and keep his lease sales punctual. And there, it turned out, was the rub. Cecil Andrus the political progressive went eyeball to eyeball with Cecil Andrus the man who takes exceptional pride in doing what he says he is going to do. Doug Foy and Francis

Bellotti and a good many others in New England started worrying about what would come first, the amendments to the Outer Continental Shelf Lands Act or Outer Continental Shelf Lease Sale Number Forty-two — for which read Georges Bank.

New England already knew a good deal about Sale Forty-two, or at least the federal government's perception of the sale, the year before Andrus arrived at Interior. In 1976 the department had released the draft environmental impact statement for the sale — the preliminary step required by law of all such federal messings with Mother Nature. By the time the statement had passed through its cycle of revisions, it had grown to more than two thousand pages of material, presented in four volumes plus a large manila envelope full of visuals — maps and grids showing fish spawning grounds, recreational areas, types of bottom sediments, circulation patterns from the bottom of the water column to the top. Spread across the four-color glossy sheets were the isobaths of Georges Bank, the prints of the great thumb. The statement, if you could take a week off to read it, was a biography of a geography. Its assemblers at the Bureau of Land Management had worked up scenarios of how oil and gas, if found, might be lifted and transported. To begin with, they said, a maximum of six exploratory rigs might drill as many as 182 holes in Georges. If they struck the "high find," the biggest field the United States Geological Survey, in its most uninhibited imaginings, could dream of, there would be twenty-eight huge production platforms, each able to drill and produce from a number of wells, out in that wild water. Production would begin roughly six years after the sale, peak in fourteen, and drop below profitable levels in twenty or so.

The statement predicted no oil spills to speak of during exploration, since none of greater than fifty barrels had been reported anywhere in U.S. waters since Santa Barbara. During the production phase, at least one spill of

37,500 barrels would probably occur (a barrel contains forty-two gallons of oil). Tankering of the oil to refineries in New Jersey might contribute another 50,000 in outright spillage along with a lot more in the form of oil-contaminated slops from tank cleaning. Other additives to Georges and nearby waters included 730,000 tons of drill cuttings from the wells and more than 230,000 tons of drilling muds.

Until they read of the encyclopedic impact statement, most New Englanders had worried in the relative comfort of the conditional. "What *would* happen, really, if those rigs ever showed up out there?" they had asked each other. Now they had some answers — lists of rigs, tankers, supply boats, spills. Now they knew. Now they spoke their minds, the most vocal of them at the hearings the Bureau of Land Management held on the statement. Some, like those who would build gas pipelines if gas were found under Georges, argued that the statement didn't say enough about the benefits drilling would bring to New England. Most others talked about the overall effects of drilling for and producing petroleum in the middle of a great fishery. Wait, they said. Wait, said the Conservation Law Foundation, until scientists can finish studying the effects of that recent spill near Georges, the one from the tanker *Argo Merchant* that drove aground on Nantucket Shoals. Wait, said Massachusetts environmental officials, until the scientific studies the federal government is sponsoring of sea life and ocean circulation out on Georges can be completed and evaluated. What's the harm in waiting?

That argument more than any other can turn a federal manager into a roman candle. Frank Basile is no exception. He has the short fuse and packed explosives of a true New Yorker, and if you want to see the show, just mention holding up a federal schedule until science says it has completed its investigation. Especially Basile's schedule. Frank is the head of the New York office of the Bureau of Land Management. As such, he was responsible for the mechanics of putting together the impact statement for Lease Sale Forty-two. If things went as he hoped they would, he would be responsible for the mechanics of the lease sale itself — opening and announcing bids, that sort of thing.

I lit the fuse when I saw Frank in his office high in the Federal Building to talk about the recent hearings on his screed. Boom. "If you don't accept prudent risks, you'd never cross a New York street," he said. "You'd sit in your room shivering." Frank sat between me and the big window overlooking the East River. All I could see of him was a silhouette, smoke curling from the nap of his natty beard. Ah, just the exhaust from a cigarette I hadn't noticed. "In this business," Frank said, "we have to say 'Tuesday is the day.' If we don't do it this Tuesday, there'll be more reasons for delay next Tuesday." He leaked more smoke. "Somebody can always say 'Stop! Research in progress!' The answer to that would be for us to stop investing four or five million dollars in those Georges Bank studies." He slammed his desk with the flat of his hand. "That'd kill that argument." Then, the explosion over, he said of course his agency wasn't about to do that.

The impetus of federal movement is awesome. An offshore lease sale gathers momentum as it moves through all its stages: from long-range planning to determining industry interest in specific sites to the multimonth effort of the draft impact statement and its public hearings and revisions, to the proposed notice of sale, to the final notice of sale, to the sale. By comparison, the wake of the amendments to the Outer Continental Shelf Lands Act as they churned through Congress was puny indeed. As

1977 wore on, Doug Foy and his allies in Massachusetts found it increasingly hard to believe that the new legislation, with its gifts to the states of a greater voice in offshore decisions, would be in place before Frank Basile stepped to a microphone in some hotel conference room and began renting drilling rights on Georges Bank.

Cecil Andrus tried to keep his lines open to the New England states. He sent governors there copies of his proposed notice of the Georges sale, something that was provided for in the language of the amendments before Congress but not yet binding on him. Massachusetts studied the notice and argued strenuously for tract deletions and more precautionary steps — such as writing some of the safeguards contained in the offshore amendments into the stipulations, the operational restrictions Interior was inserting in leases for the more biologically sensitive tracts.

Doug Foy was not pleased. All these federal-state negotiations, it seemed to him, were obscuring the fact that the amendments remained unlegislated. Correcting mistakes by adding stipulations would be difficult once the leases were signed, he said. And enforcing all those regulations would be next to impossible, particularly in the fury of the offshore winters, when inspectors would have even fewer opportunities to get out to the rigs.

In the late fall of 1977, the give went out of the dickering between Boston and Washington. Cecil Andrus, Democrat, of Interior, could not guarantee Governor Michael Dukakis, Democrat, of Massachusetts, that the United States Congress would pass the legislation both men wanted.

"It was the beginning of a very interesting and compressed war," Doug told me one summer day in 1982, seven years after the fight for Georges began. We sat in his office, he stealing a look now and then out his window at the summer afternoon drifting along the bricks and shade of Joy Street, and I darting envious glances at his racing bike, parked in front of the fireplace. Andrus would seem to be signalling that he might delay the sale, Doug said, and then he would appear to reverse himself. But on the day of New Year's Eve, the secretary had put his foot down.. Lease Sale Forty-two, Andrus declared, would be held at the end of January 1978 — in one month.

"There was some doubt that Governor Dukakis would stick it out and litigate," Doug said. "My understanding is he was getting a lot of advice not to." Attorney General Bellotti seemed firmly in support of a suit, though, and Dukakis came down on his side. When Doug Foy heard about Andrus' announcement of the sale date, he was at his parents' house for the holidays. "Okay," Doug remembered saying to himself, "we'll be in court in a couple of weeks."

The law, realizing that it often has trouble in a foot race with a snail, has made a provision for those who need quick relief. This procedural Rolaid is called the preliminary injunction, and it enables the successful plaintiff to stop an action that allegedly would cause pain to said plaintiff until the court system can catch up and look at the merits of the case. If you wish to get a preliminary injunction, you must appear before a judge (in offshore matters you need a federal judge) and show that you have a case whose merits are likely to prevail; that you will suffer irreparable harm in the absence of the injunction; that the injunction will not harm others; that it will be in the public interest — things of that sort. Not easy, particularly if the action you seek to block has just been blessed by the head of one of the most well-muscled agencies in Washington.

The commonwealth and the Conservation Law Foundation filed suit in the middle of January 1978. Everyone trooped over to the post office building in the funky laby-rinth of downtown Boston. And there, presiding over the Federal District Court for the District of Massachusetts, was the Honorable W. Arthur Garrity, the busing judge, the man who caught America's attention by directing the integration of Boston's schools from his bench.

A herd of litigants filled Garrity's courtoom: Depart-ment of Justice lawyers representing Cecil Andrus and Juanita Kreps, the Secretary of Commerce (the plaintiffs were suing her too because they said she was letting Andrus have his way with Georges Bank and thus not doing her job as manager of its fisheries); people from Bellotti's office representing aggrieved parties like the state Secretary of Environmental Affairs; lawyers for the defendant-intervenors — companies, like Arco, Mobil, and Exxon, that were ready for Lease Sale Forty-two and wanted no monkey business; and Douglas Irving Foy and a couple of his colleagues. Plus press, fishermen, preserv-ers of Cape Cod, defenders of beaches, partisans of whales.

Documents for the record, including Frank Basile's tomes, mounted in columns. Judge Garrity listened to three days of arguments. Bellotti's lawyers complained that defendant Andrus was not paying enough attention to Massachusetts' fishing and tourist industries; that he had violated fishing rights by not paying attention to the plans then being developed by the New England Re-gional Fishery Management Council, a body recently es-tablished by Congress to manage the cod and haddock and scallops and other tenants of Georges Bank; that he had violated the sacred and seminal National Environ-mental Policy Act (NEPA) by not properly balancing en-vironmental risk against economic benefit when he decided to lease Georges; that he had violated the Endan-gered Species Act by not giving sufficient weight to the hazards that might befall whales cruising in the drilling areas of the Bank.

Foy's brief took pains to point out that "plaintiffs do not seek to prevent offshore oil and gas leases from ever occurring in the North Atlantic. Federal laws exist to promote harmonious development of living and non-living resources, but defendants haven't complied with their respective statutory and common-law obligations to conserve and manage marine fisheries resources."

"I stood before Judge Garrity," Doug recalled on that summer day in 1982, eyeing the great outdoors of Beacon Hill, "and I said that except in certain limited respects this was not a NEPA case." Contrary to the commonwealth's arguments, Foy was contending that the suit did not centrally involve the National Environmental Policy Act. "What it was," Doug said, "was a duty case — the duty to protect the Georges fishery." Juanita Kreps had that duty, Doug insisted, under the same federal law that created the regional fisheries management councils. Andrus also had the duty, under common law and under statutes like the old Outer Continental Shelf Lands Act. "You couldn't fly in the face of that with a separate federal program that would in theory have the potential for sacrificing the fisheries; you couldn't go blasting in with an oil and gas program and say 'fisheries be damned.' "

Foy laid another complaint on Juanita Kreps: the summer before the suit, the National Marine Fisheries Service had nominated Georges as a marine sanctuary, a step that would preserve the Bank for conservation and ecological research and, in effect, give the existing fish resource the edge over putative oil and gas resources. Foy said that the Secretary of Commerce, who must pass on such nominations and approve those that are sent on to the President, hadn't completed her deliberations on Georges yet. If the leases were sold and the rigs went in, he argued, there wouldn't be any reason for her to do so; the rights of the leaseholders, the oil companies, would then have to take precedence over the desire to preserve.

Oh, come on, said the lawyers for the federal government and the oil companies. These environmentalists never change their tune. They're always trying to bring up obfuscatory issues. The central issue here, Your Honor, is the environmental impact statement. The ones for earlier sales may have been too short or too vague, but this one made full use of the latest technological information available when it was prepared. Also, Secretary Andrus has deleted from the sale all the tracts Massachusetts wanted deleted, and he has added protective stipulations to all those that are left. Remember, Your Honor, the Outer Continental Shelf Lands Act reminds us all of the urgent need for further exploration and development of the outer continental shelf. And finally, we protest the plaintiffs' tactics in delaying until the last minute to bring suit; they are trying to force an injunction on the court. If they're successful, a lot of companies that have invested large sums in bid preparation, rig contracts, and the like, will suffer.

Three days before sale date, Judge Garrity released his decision. Garrity said he would grant a preliminary injunction — not against the sale but against the reception of bids for the sale. No bids, no deposits. No deposits, no harm done to the oil companies.

"This is no ordinary fishing ground," said Garrity. "It is as important a resource as the people of this state will ever have to rely upon." It would be a violation of Cecil Andrus' duty, Garrity said, for him to receive bids on Georges while amendments to the Outer Continental Shelf Lands Act were pending in Congress. "The whole point, as I understand it, of the secretary seeking this legislation . . . is premised on his conclusion that he lacks power to accomplish these ends administratively." The defense, the judge said, had insisted that the sale should proceed because in due time Congress would pass the amendments. The question, Garrity said, should be if Congress would indeed pass them, and if they would be retroactive to Lease Sale Forty-two. "The Court's response is that the big word in that argument is the smallest . . . *if.*"

Garrity found the sale's environmental impact statement long on identification and short on evaluation. It didn't consider the costs and benefits of delay. It didn't address the alternative of making Georges a marine sanctuary. It did not dwell at any length on possible damage to beaches on Cape Cod and the islands. Andrus, Garrity said, had been dilatory in deciding whether to hold or postpone the sale. The judge quoted Andrus during a news conference a couple of weeks *after* the secretary had announced the sale; Andrus had said he still had made no decision on whether to wait for the amendments he wanted from Congress or not. "When the secretary vacillates in the fashion exhibited here without explaining the reason therefor, it is, in my opinion, open to the court to conclude . . . consistently with the rule of reason that's required in these situations, that the action is arbitrary and capricious."

Plaintiffs would suffer irreparable harm, said Garrity, since Georges could never become a marine sanctuary once drilling started. "The plaintiffs are looking for a delay of a relatively few months to preserve a resource that has taken millions of years to accrue, and which will be with us, for better or for worse, for untold centuries to come. . . . The opposing considerations here are use for a period of about twenty years as a source for gas and oil, as against the preservation of the natural resource . . . for the indefinite future." The judge then summed up his summation: "If there ever was a public interest case, this is it."

That was that, for the moment, except for some activity on the defense side, which announced it would appeal and made a pro forma motion that appeared to irk the departing jurist a trifle.

The defense: "Let me present the motion for stay, Your

Honor, just for the appellate court, so that it is clear a motion was made here."

The court: "What is the motion — to stay what?"

The defense: "To stay the effect of your preliminary order."

The court: "My order was to the effect of staying the secretary's action, so how can I grant a stay of my order to stay the secretary's action?"

The government and the oil companies rushed to Judge Levin H. Campbell of the United States Court of Appeals for the First Circuit, also in Boston. Well, said Judge Campbell, an emergency stay of this injunction would be possible only if it can be shown that the district court acted beyond its authority. Campbell thought that the "judge below" had not done so, that Garrity had not, as the government claimed, invaded the powers of the executive branch, that Andrus possibly had acted capriciously by deciding he would accept bids before Congress acted. Then the judge said, "There may be issues more serious than ones involving the future of the oceans of our planet and the life within them, but surely they are few."

Doug Foy framed a copy of Campbell's ruling and hung it on his office wall. That was the point, he wrote me as I was finishing this book. Balancing the resources was the point. And to do that you had to force Interior, "which considers its role primarily that of a leasing and mineral exploitation agent," to restructure. He hadn't gone to court to delay the lease sale, though delay admittedly was a tool in his bag, but to try to apply pressure on Interior, through the court, to reexamine its perception of mission. What he wanted Andrus and his people to ask themselves was: "Are we going to develop oil and gas wherever possible and protect fish as an afterthought, or are we going to husband our existing resources and develop new ones, such as oil, when that development does not threaten those existing resources?"

With the stay denied, Cecil Andrus decided to wait for Congress and let the appeal run its course in Campbell's court. On January 30, with one day to go, he cancelled Lease Sale Forty-two — for the time being.

Fishermen's wives are unlike other wives. Farmers' wives are partners, but they are likely to be partners in place. Many offshore fishermen's wives spend more time alone than with their husbands. They must look after the family, the community, their husbands' interests, their interests. They have learned to work together, and nowhere better than in Gloucester, the old and now Italianate port northeast of Boston. The Gloucester Fishermen's Wives Association is a force not just along the docks but along the Potomac River. When government people who know New England start thinking of reaction to their plans for Georges Bank, they think early on of the Gloucester wives and, in particular, of Angela Sanfilippo.

She is a small, sweet woman, Angela, young — around thirty, I'd guess — and strong. Doug Foy says she really feels the history of Gloucester, which might be strange, since her motherland is Italy. But Angela has an extraordinary commitment to place and cause. The Catholic religion is strong in the Gloucester fleet and among the Gloucester wives, and it shines in Angela. Georges Bank and its preservation as a fishery are articles of her faith.

In New England, fishermen don't account for more than a couple of percentage points in the work force and not much more in its total economic output. But they are concentrated in the port towns and they are a contentious constituency. They are food producers, and that still means something, even in a society only vaguely aware of how it is fed. They are seafarers. "If the fisheries of New England are no longer the bases for our wealth," wrote Samuel Eliot Morison, "if we have turned inland and inward from the coast, eat less fish than we once did and are unconcerned with or unconscious of our historical and cultural roots, even if all these things are true, fishing boats and fishing ports retain their hold on our native imagination." Perhaps Congress has felt that hold. Or perhaps it is mostly horse-trading in the cloakrooms that gives Yankee fishermen their edge. What is certain is that the fishermen have learned to trade hard on what they represent. Protesting farmers drive tractors as big as earth movers around Washington. Protesting fishermen send an occasional boat up the Potomac or haul a dory black with signatures to the White House. Boat owners now know the Hill about as well as they know Cultivator Shoals.

So do their wives. Angela testified before Congress. She spoke in her high, husky voice, no written statement in front of her, about what she knows and believes, about the bravery of her fishermen, about "the food of fish, the only natural protein that is left in the world." She was, she said, "a fisherman's wife and a Gloucester person," and she wanted Congress to see to it that her husband, John, a trawler skipper, and the rest of her people would continue feeding the world from Georges. The chairman of the Senate panel, Paul Tsongas of Angela's home state, said, "I would hate to run against you after that."

Angela lives near the main road to Boston, thirty miles along the coast. The neatness of her house is startling. Her kitchen is so clean everything looks as if it has been coated with clear plastic. She sat in it one warm afternoon in the spring of 1982, talking about her work and how she met Doug Foy. She had a cold, and her little daughter in the next room had one too. Her child's name is Giovanna. If she had been a boy, Angela said, her name would have been George.

In the late seventies, the fishermen's wives began attending meetings on matters that might affect their husbands, their fleet. They spent a lot of time at meetings of the New England Regional Fishery Management Council, listening to the confusion, gradually learning about quotas and closed areas and fishing mortality. And about the idea of a marine sanctuary for Georges Bank. The council took up a proposal on the matter from the National Marine Fisheries Service. "I never heard the word *sanctuary* before," Angela said. She and her friend Lena Novello asked around. "Somebody told us that a marine sanctuary means you can't fish there no more, and suddenly we got scared. We went and opposed the whole thing." Fisheries backed off.

Angela said she first met Doug Foy when he was working on his brief for the suit he took to Judge Garrity's court. By then, the fishermen's wives had a name along the coast, and Doug wanted Angela to go along and talk with Governor Dukakis about Georges Bank. "Then the whole thing opened up," she said. "Lena told me, 'Maybe there is something to this marine sanctuary.' We didn't know, we were fooled about it." The Fisheries people finally convinced Lena and Angela that a sanctuary wouldn't preclude fishing but would regulate drilling. In the spring of 1979, when the Conservation Law Foundation sent one of its lawyers to Gloucester, the wives told her they might be willing to back a sanctuary. "Those CLF people were fantastic," Angela said. "They did the legal work, we did the practical work." Under the law, sanctuaries can be nominated by private citizens and groups as well as by agencies, and that was the work Angela and company undertook. "It was in May of 1979," she said, "when along with CLF and the New Bedford people we sent the Secretary of Commerce the information that we wanted Georges Bank to be a marine sanctuary."

Shortly after talking with Angela, I asked Doug Foy about working with her. "We were aware that a marine sanctuary program should be considered," he said, and his lawyerly tone clashed with my memory of Angela's energetic artlessness. "It was a way to flip the presumption that you'll do oil and gas primarily and try to protect the fish secondarily. You get the primary functional authority and control in a fisheries agency as opposed to an oil-and-gas extractive agency." The National Marine Fisheries Service didn't do its homework when it first proposed the idea for Georges Bank, he said. "The fishermen went crazy. They thought it was going to be a bird sanctuary." When Doug and his people talked to the government about trying again, "they said if you get the fishermen to go along, we're willing to go ahead." With Angela's help, the Conservation Law Foundation got the support it needed.

Meanwhile, Cecil Andrus' plans for the North Atlantic were back on stream. Congress passed the Outer Continental Shelf Lands Act amendments, giving New England just about everything it had asked for to protect Georges Bank. In February of 1979, Judge Campbell vacated the injunction against Lease Sale Forty-two, saying that the amendments rendered the central argument for the injunction moot. Campbell found that Congress clearly believed it was possible and desirable for both fish and oil to be taken from Georges Bank, with proper safeguards. But he felt Andrus had not paid enough attention to the sanctuary alternative. "Should there be particular areas of Georges Bank that are uniquely important to the fishery," he wrote, "the management by the Secretary of Commerce, the administrator of the Fishery [Conservation and Management Act], rather than by the Secretary of the Interior might be advantageous." In response, Interior prepared yet another revision of its environmental impact statement, a supplement that took in both the sanctuary issue and the new amendments.

Nineteen seventy-nine was a year of gubernatorial change in New England. In Massachusetts the conservative Edward King had succeeded the liberal Michael Dukakis. The mood in the region turned cautionary. Many politicians there thought the sanctuary idea wasn't far enough along to warrant discussion one way or another, and in any case Congress had now provided enough protection for Georges so that Forty-two should go through as planned.

Doug Foy, the Sierra Club, the Natural Resource Defense Council all argued for delay of the sale, saying the regulations implementing the new amendments hadn't been put in place and that things should be held up until more was known about the effects of drilling muds and other discharges on the fisheries.

The sanctuary proposal Doug and Angela had sent to Washington began to stretch tempers in high places. The National Oceanic and Atmospheric Administration, the agency in charge of the sanctuary program, held hearings on it in the summer of 1979. In September, Interior and Commerce announced there would be no sanctuary on the Bank. Rather, there would be a biological task force monitoring drilling operations there. The task force would include representatives of the government agencies concerned, the affected states, and the New England Regional Fishery Management Council.

"The people from Washington came up," Angela told me in her kitchen. "We met with them, in this house right here. Then out of the blue, they come out with the announcement. The excuse they made is they had some deal with the Department of Interior and the whole thing was being dropped. I will never forget that day. It was like, you know, when somebody dies in your family, people come and visit you? People called from all over the country. People called from Alaska."

What happened? To find out, I consulted Robert Knecht, head of the coastal zone program in the National Oceanic and Atmospheric Administration — Fisheries' parent agency — who had run the hearings on the sanctuary proposal in New England. He felt there wasn't enough solid support for it from the fishermen, at least from the Fishery Council. "They saw it as more federal authority," he says, "and they wanted less." The entire marine sanctuary program was young and tender; a bill was already pending in Congress to kill it. "We could lose if we made a mistake," Knecht says. "To use a sanctuary to manage oil and gas seemed to me increasingly wrong. That affected me a great deal, and I think that in the end the decision on the Georges sanctuary was mine."

There are stories that NOAA used the sanctuary merely as a threat to get the best deal it could with its far more powerful sister agency, Interior; that the Carter White House, anxious not to have its prized offshore amendments diluted by the competing claims of a sanctuary on Georges, put in a phone call or two to NOAA.

"Absolute power politics," Doug Foy says. "Interior tore NOAA's face off. It told them to stay out of oil and gas. Things like the biological task force were just to save face. If Interior had not been the lessor, if it had been J. Paul Getty who owned Georges Bank and was going to lease it, I bet you a thousand to one NOAA would have sued him."

The Gloucester fishermen's wives went to Washington to protest. "I couldn't go," Angela says, "because I was in my last days of having Giovanna." Doug Foy went back to court.

Round two, ladies and gentlemen. There's the bell.

First District Court, Boston. Conservation Law Foundation, plaintiff, along with the state of Maine and the commonwealth of Massachusetts — or part of it: Attorney General Bellotti joined with Foy, but Governor King filed an amicus curiae brief on the other side.The arguments were pretty much the same as in round one, with the additional complaint by the plaintiffs that the Secretary of Commerce and her officers had been "arbitrary and capricious" in withdrawing the sanctuary nomination. The decision was very different. Judge J. J. McNaught found no evidence that the plaintiffs would suffer irreparable harm. "Indeed," he wrote, "physical harm to the environment would not occur until the exploratory drilling began." Regulations covering such safeguards as the use of best and safest technology had not yet been written, McNaught noted, but "I have no reason not to believe that the secretary will impose . . . those requirements subsequent to the sale. Continuous control appears to be part of the lease terms." Nor did plaintiffs appear likely to prevail on the merits of their arguments. Motion for injunctive relief denied.

Foy and Bellotti appealed, and lost again. The appeals court did stay its ruling so that the losers could get to the Supreme Court. As Frank Basile and a crowd of people from the oil patch started to gather in Providence, Rhode Island, to begin the sale, Foy and a couple of Bellotti's lawyers scrambled to get to Washington. Doug remembers running uphill from the appeals court back to his office, grabbing some papers, grabbing a taxi, and sprinting for the plane. At the Supreme Court, an obliging clerk found him a manual typewriter and a cubby in a back file room. Foy and his tiny team filed their papers with the clerk, who whisked them off down the marble corridors to Justice William Brennan, who had the helm on the case.

"Somewhere around four o'clock," Doug says, "comes a stay out of the chambers of the Court. Then all hell broke loose. Within ten or fifteen minutes, the Solicitor General of the United States and literally twenty attorneys for the oil companies and the government roared into the Supreme Court, and the Solicitor General filed an immediate request for a rehearing." The cubbyhole gang used the Court library as ready reference to help in typing up responses to the government pleadings and sent them off down the corridors. Rehearing request denied. The oil people tried on their own, sending pleas for a rehearing, handwritten on yellow legal cap, to the Chief Justice, who had gone home for supper. At one point, Foy and an old friend representing the oil companies were standing in the corridor together. The friend, who was charging several hundred dollars an hour for his services at that moment, clutched his pen and legal pad. "I pointed out," Doug said, "that the way to tell his side from our side was that we could type."

In the Grand Ballroom of the Biltmore Hotel in Providence, a couple of hundred senior employees of some of the most powerful corporations in America sat and sat. Then Frank Basile got the word from Washington. He went to a microphone and told the oil men what had happened. "This," he said, every word an icicle, "is the American system of justice in action." Someone clapped, slowly, three times. Someone roared "Bullshit!" Then the crowd rose and, deltoid boot tips blinking in the television lights, broke for the phones to change reservations back to Houston.

Brennan's stay was only until the full Court could review the case. The Court vacated the stay, but by then, bureaucracy being bureaucracy, a new notice of sale had to be issued with another thirty-day lead time. And because of a typo in that, it wasn't until December 18, 1979, that Frank Basile got back to his rostrum.

Even as renovated, the Biltmore Hotel in Providence, Rhode Island, has the look of the lorn. It is a relict from the better days of train service in and out of the fancy station close by up the hill and reliable commerce downtown. Inside, it is almost foolishly cheery. The Grand Ballroom on the seventeenth floor is done in pinkish plaster, and brown cherubs bulge from the ceiling. In the big bar down the hall, a manager calls to his people polishing tables along the window wall. "One hour," he says. "One hour and we'll be up to our eyeballs in oil men."

The floor of the Grand Ballroom, this week before Christmas of 1982, looks like a warehouse for old collapsible chairs. Television crewmen stand hipshot along the walls, their cameras perched like buzzards on their shoulders. It's early, and this is Providence, but something might happen. This is the first time they're offering oil and gas leases on such an important American fishing ground. Oil versus cod. You know, Texas versus the Yankees. Something might happen.

Frank Basile is arranging papers, talking to people in the crowd. Short, dark suit, dark beard, natty. He has been around. He was one of the first in the door when the Bureau of Land Management opened the New York office in 1973. Before that he was a wildlife management biologist and before that an oceanographer. Remembering what happened the last time under the pink plaster, he is eager to get on with the sale. A reporter has picked up a rumor that the commonwealth is trying for another stay. "Here we go," the reporter says.

But we don't. In the forenoon, Basile announces that the sealed bids the companies have submitted for tracts in the sale will be opened after lunch in an auditorium nearby. "The Department of the Interior has gone to extremes to allow the judicial system to work," he says. "We've tried hard to provide safeguards wanted by the plaintiffs." He grins. "Now we're not going to answer the phones for a while."

At one o'clock, the sale begins. The oil companies have submitted 189 bids for 79 of the 116 tracts being offered. Basile's assistant sits beside him up on the auditorium stage; he opens a large manila envelope and passes the bid over to his boss. "Tract ten," Frank says. Paper rustles as the audience finds tract ten on its maps, way up in the shallows above the head of Oceanographer Canyon. "The only bid is from Pan Energy Resources, Challenger Minerals, and Oxoco; two hundred thousand six hundred and sixty-six dollars." That is pin money in this kind of betting. Another bid. Two partners, Conoco and Getty, are interested in tract sixteen, east of tract ten, also in shallow water. But Tenneco wants this one, wants the right to drill on this three-by-three-mile square of ocean bottom for the next five years, renewable, and wins it for $1,619,000. Things are picking up. Basile reaches for another bid.

Suddenly, the sound of fluttering paper and of heavy liquid hitting the floor. A handful of young people are standing over us in the balcony, dumping boxes of pastel paper over the railing and throwing plastic bags filled with something black. Oil. The stuff begins to ooze under the seats. One bag hits a man in the side of the neck, blackening his right side. He stands up and, calm in his shock, slowly takes off his coat. Television crews run for him like chickens after a sick hen.

The message on the sheets reads, "Today, December eighteenth, here in Providence, Rhode Island, the United States government . . . and the court system have once again taken it upon themselves to auction off . . . Georges Bank, one of the world's richest fishing grounds, to the multinational oil industry." It accuses Cecil Andrus and the oil companies of deceit and worse in placing the dollar "before the benefit and welfare of the common folk." The authors, who call themselves Fish for Survival, demand that leasing stop until more scientific study is done on oil and fish. "The oil spilled today," they say, "is a drop in the bucket compared to what will happen if the oil companies get their way."

Down at the bottom of the page is "ON THE OCEAN THERE IS NO ESCAPE FROM DISASTER." In the balcony, though, there is. Fish for Survival wriggles past the police.

On the floor, people move to the back of the auditorium away from the oil and the stench. But Frank Basile asks them to put up with it, and they come, stepping gingerly over the spatters and slicks of their product, back to the game. As they settle in their seats, Joseph Garrahy, the Governor of Rhode Island, hurries in to apologize. He is gloriously righteous. "We don't condone lawlessness in Rhode Island," he says. "We're delighted to have you here, we hope to be self-sufficient in energy some day." He also hopes that offshore drilling off New England, which will use old Navy facilities in his state, will help his economy, but he doesn't say that.

Basile vents a little. "Always nice to see the governor," he says. "One of the few people in New England who says nice things about us."

Patterns begin to appear in the bidding. The most striking is that of the Mobil Oil Corporation. Three hundred oil companies could be represented here if they wished; they have the requisite experience and financial standing to satisfy federal standards. Thirty-one companies are in fact participating, including Exxon, Shell, Mobil, and Gulf. By law, these and other companies with a worldwide production of more than 1.6 million barrels a day of major petroleum products are banned from bidding together, but the giants can and do consort with those of more mortal dimensions. Mobil has come to Providence with Tenneco, Amerada Hess, and Transco. It has come to win.

For months and years, most of the companies present have been putting the spurs to their geologists, geophysicists, seismologists, and technicians with expertise too specialized to name, trying to come up with the best guess on what lies under the grid of tracts on Georges Bank and how much to bet that it's there. So far, the entire Atlantic continental shelf of the United States has produced nothing in commercial quantities. Out of the twenty wells drilled so far on Baltimore Canyon, sixteen have been dusters, dry holes, and three have produced gas or oil but not enough. So there is little precedent to go on.

"It's like playing Battleship," one oil company head told me. "You're trying to figure out if the treasure lies under one block or a string of them."

Exxon, the giant of giants, always bidding alone, takes the early play. You can see how subjective, how quirky, the bidding strategies are. One of the games the bidders play is called money left on the table. It is how much you lost by winning — the difference between the top bid and the runner-up. Atlantic Richfield submits a bid of $2,650,000 for one tract. Phillips and its group bid $3,333,000, Chevron and its partners $4,221,000. Then the Mobil group pounces: $25,137,000. Wrong, says Exxon: $51,797,000. A lot more than earth-science estimates goes into that spread. Some companies bid the way they do for financial reasons or to trip up competitors. Some, like Mobil, are low in domestic reserves and need to find new supplies. Some, like Mobil, have strong ties to the East Coast. They feel, my Battleship man said, that "if there's going to be a big hydrocarbon discovery out here on Georges Bank, their name had damn well better be in the pot."

But Mobil is operating on something more motivational than general exploration strategies or pride in its New York beginnings. Money left on the table is no object. The Mobil group puts down $75,238,000 for one tract; the runners-up, the Shell group, offer only $3,133,000. Mobil climbs to $80,299,000 for another, the high bid of the day, against $13,175,000 from the Shell group, and to $79,196,000 against $12,137,000 from Exxon.

When Basile reads the last bid, the pattern is clear. The Mobil partners in their various combinations have put down half of the high-bid total. They have bought drilling rights on sixteen blocks (tracts tend to become blocks after they have been leased). They are connected, in a winding way, in a line centered just inshore of the head of Lydonia Canyon and lying roughly east and west along the fifty-fathom depth contour or isobath. Whatever Mobil was after, it seems to have got it.

Frank Basile comes down off the podium. "Excellent results," he says. "I was expecting maybe six hundred million in high bids, and we got over eight hundred. More than two bids per tract. That's good." No records here, but a satisfactory return for the federal government, which depends upon oil explorations on public lands as its second largest source of revenue after the income tax.

A young woman pushes through the group around Frank. She is angry, hunching forward. "Don't you feel," she says loudly and moves closer, "don't you feel that when you're selling these leases you're really selling off Mother Earth?"

It's almost as if Basile has been waiting for the question. "No more so," he says, "than when I eat steak."

4

Mobil's Reef

A LITTLE after midnight in late July, 1981, the semi-submersible drilling rig *Zapata Saratoga* made a noise like a great bell rolling over cobbles. It was the sound of a drill stabbing the sediments of Georges Bank at a point 155 miles southeast of Nantucket: Block 410, leased by the Shell Oil Company and its partners at Outer Continental Shelf Lease Sale Number Forty-two with a high bonus bid of $34,733,000. Things did not start out well for *Saratoga*. On her arrival a couple of weeks before, she had squatted down close to some buoys marking the traps being worked by a lobster boat out of Hyannis on Cape Cod, and the captain had howled. Someone phoned in a bomb threat. Representatives of a particularly action-oriented group called Greenpeace picketed the rigs in an assortment of craft, bearing signs reading "Oil and Water Don't Mix" and "Food Not Crude." The worst insult came from Georges Bank itself. The anchors wouldn't hold. Finally, the rig had to be towed in for refitting with new gear. For all that, *Saratoga* made history, at least in the New England press. She was the first to penetrate the Yankee offshore frontier. *Alaskan Star*, working for Exxon, was second by twelve hours.

It is November of 1981 now, just before Thanksgiving, and the dreary wind bellies over the runways and deep-water docks that served the country so well forty years ago. Quonset was a familiar name to sailors and flyers of World War Two, and for a time after that, bases here and elsewhere in Rhode Island made the Navy the state's largest employer. But the Navy decamped abruptly, in 1973, and the population around Quonset Point and nearby Davisville dropped by a third. Now the region, the historic home of the Swamp Yankee, the loner of loners, lies desolate as the cracked concrete.

At Davisville, the state has hopes for an industrial park to bring its South County back. You'd never know it to look at the place, but about eighty companies doing oil-related business accepted Governor Garrahy's invitation to come on up. Davisville has become the main supply base for drilling from Baltimore Canyon north. The Mobil office is on the second floor of Building Number Seven. A mile or so away are the docks it will use to move supplies to Georges Bank.

In a way, Georges is as foreign to Mobil as parts of off-shore Africa or Indonesia. If a field is found, Mobil usually sets up a subsidiary to handle it. Until then, the work is done or overseen by a group of world travellers, including the employees of MEPSI — Mobil Exploration and Producing Services, Inc. — based in Dallas. "They work without regard to political boundaries," the MEPSI chief of exploration is quoted as saying in a company newsletter. "The concept is that hydrocarbons exist independent of political considerations, or lines drawn on a map."

Al Mitchell is with MEPSI. He is gentle and pleasant. His business is positioning rigs. He has lived in Paris and the Hague and has placed rigs where MEPSI says they should be placed, from offshore Honduras to the Red Sea and on around the world. His last job, just a month or so ago, was in Egypt. "I tell 'em to worry about the well vertically and let me worry about it horizontally," he says. Al has been with Mobil twenty-five years.

Al and I are going out in a few hours with a couple of surveyors aboard the supply boat *Millie Bruce*, just up from the Gulf of Mexico. *Millie Bruce* will set anchor markers for *Rowan Midland*, the rig Mobil has hired for its work on Georges. *Midland* will drill two wells, possibly a third. She is now under tow by two tugs out of New Orleans and she has been hitting heavy weather — twenty-foot seas and sixty-mile-an-hour winds. This is Tuesday. The rig is due on Georges Saturday.

Millie Bruce is Gulf design, exhaust stacks low on either side of a work deck long enough to carry pipe or a covey of office vans full of equipment for the outfits that run tests on the well or handle the drilling muds. Right now, eight rusted anchors are secured in pairs athwartships. They are over twenty feet long, made by Vicinay in Spain. The words on the ginger iron say that and "30,000 Light Weight." Fifteen tons apiece.

These are piggybacks. *Midland* carries eight roughly similar anchors in stanchions low on her corner legs. The workboats will carry those out four thousand feet from the rig and set them, according to the plan Al has, and then the Vicinays will be attached to wire running from those primary anchors, run out another five hundred feet, and dropped. Big, bright can buoys will mark them, attached by cable strong enough to break the monsters free of the bottom when the time comes to move to another block.

On the walkway aft of the bridge, Al shows me a sheaf of bamboo poles stuck through plastic floats and rigged with red-and-white flags. They'll mark the spots where the anchor boats are supposed to drop their loads. Al flips some papers. "This well," he says, "will be eleven hundred and twenty-one meters from the north line of Block Three-twelve and eleven hundred and sixty-nine meters from the east line." He reads out longitude and latitude, down to hundredths of seconds. Most of the work will be done using a shore-based navigational system that can be accurate to two meters, though the stations are going to be awfully far away, Al says, for this job. Final positioning, by rule of the United States Geological Survey, or that part of it that oversees drilling operations offshore, will be by satellite, and that is accurate to one meter. The surveyors are up in the wheelhouse, three of them. They talk in units of Universal Transverse Mercator, placing *Midland*'s position so many hundred thousand meters east of a central meridian and so many million meters north of the equator. They are all employees of John E. Chance, one of the most precise of positioners. They drove up to Davisville from a job in the Gulf, on their way to a job somewhere else. Like so many in the oil patch, they are migrants with a single skill, the world's most sophisticated tinkers.

Kurt, one of the surveyors, goes over the bearing of the anchored rig with Al. The bow will head 310 degrees. We will follow an octagonal course, dropping markers every 45 degrees. Then Kurt checks his instruments: a column of computers topped by a readout screen (a small CRT, or cathode ray tube) and lashed down with nylon line and duct tape. He runs some plastic program cards through his microcomputer. "Beautiful," he whispers to himself.

The skipper's name is McAllister, but he prefers to be called by what is stitched over the breast pocket of his tan, short-sleeved shirt. He, a Gulf man, is going to wear his Gulf shirt in this Yankee arctic right on through, even if he has to melt the wheelhouse heaters. The stitching says "Capt. Mack." Capt. Mack has a sun-pickled face based on radial wrinkles anchored on the corners of his eyes and mouth. He is, he says, "wore out from holdin' on" during all the storms they hit on the way up. It's so hot in the wheelhouse that the windows are running water. "Got to wring a little water out of her," he says, flipping the switch of a defogger. It doesn't do much good.

Narragansett Bay is calm and cold. Capt. Mack takes it easy in these unfamiliar waters, worrying himself around a wreck marked on his chart close to the mouth of the bay. He hits two lobster-pot markers, bobbing in the channel where they shouldn't be, and swears. Then we are at sea, watching the sun drop from a cloud into the horizon.

By six-thirty the next morning, we are lying off *Alaskan Star*. She has completed her first well, and Exxon has declared it a dry hole. The news was on television last night. *Alaskan Star* is tidying up before moving on to another Exxon block, but before she does, she'll provide a known location point for our surveyors. One of them climbs into an orange survival suit and waddles out to the work deck. The rig sends down its personnel basket, a rubber doughnut hung in a heavy net, spiralling at the end of the crane wire. The basket hits the deck, the surveyor shoves his instruments onto the floor of the doughnut and grabs the net. The crane operator jerks him upward like a hooked fish.

We wait for an hour to get our man back. Kurt tells me he did the positioning for *Zapata Saratoga*. A shore station malfunctioned, he says, and he thought he'd never get the fix. But the station came back on. Keep those things to a minimum, he says, and surveying is a pretty good life. "You meet just about every kind of person. Like this guy a friend of mine told me about, some Ayrab who wanted him to line up his prayer rug with Mecca a couple of times a day."

We start setting markers. The flag of the first one stands in the wind and is lost in the following sea. The others go. A surveyor stands behind Capt. Mack, guiding him to the release points, his eyes locked on the CRT screen. "You're on line," says the surveyor, but Capt.

Mack muffs it. "Damn sea's on muh stern," he says. He circles for another go and puts *Millie Bruce* right on the money.

Al comes into the wheelhouse and points out at the swells. "That's where the oil field is," he says. The currents pull the flags down until some are completely under. In a couple of hours, the tides will swing and release them. We go too close inspecting the process and cut an anchor line. We replace the marker and stand off, at anchor, in twenty-eight fathoms.

Millie Bruce has a fine galley, and that is important on workboats. Capt. Mack says he knows of a Gulf skipper who got on the radio and said he was all shut down. "His cookin' range was out and his toaster was broke." The crew are kindly, passing around the tasty beans and ham and hot sauce. But here, too, the seaman at mess, like Kipling's cat, prefers to walk by his wild lone. The meal is taken in silence. Afterward, in the wheelhouse, we hear a southern voice on the side-band radio. "I'm a little hard of hearing," says Capt. Mack. "Stayed in the engine room too long." He turns up the volume. *Midland* is talking to her tugs. She is coming in.

Saturday is blown away by the wind. The seas are topping ten feet. *Midland* asks Capt. Mack if he can raise the two anchor boats that are supposed to meet us out here and see if they can work in weather like this. Capt. Mack can't find the first one, *Ocean Ray*, but after a while he raises the second, *Seaforth Jarl*, a boat Mobil has been using up off Canada. The captain of *Jarl* is a Scot. So is McAllister, but it is some time before the Gaelic blood wins out over the mutually confusing accents and Capt. Mack is able to tell the rig that the seas are too high.

"Thanks," says *Midland*. "Thanks. 'Preciate it."

So here we sit, anchored in the wind, waiting for calm. Somewhere out in the ruck, two coonass tugs are trundling *Midland* around like someone in need of sobering.

They can't anchor her hugeness, so they walk her up and down the seas and wait. Each day costs Mobil $55,000 in rig fees, and, what with inflation and the tight rig market, that is soon going way up.

Coonass is to Cajun what *Polack* is to Pole. If everyone is friendly, you can use the word and get smiles. If not, you can use it and get your front teeth shortened. No one really knows where the term comes from. Some say from the oil fields of southeast Texas, where the Texas drillers used it to put down the cheap labor from Louisiana, the Acadians, who fled to the bayous after the British drove them out of Nova Scotia. Now Cajuns are among the masters of offshore drilling, and some of them like bumper stickers and T-shirts with coonass jokes. But among those who remember their Acadian heritage and pay homage to French culture, *coonass* is apt to be an insult. Capt. Mack says he speaks a little coonass. His wife speaks Cajun and Parisian French, he says. Right now, he's bobbing within a day's steam of the forest primeval, the murmuring pines and the hemlocks, where his wife's people — and Longfellow's Evangeline — once had their home.

Sunday the seas drop, but not enough to run anchors. *Ocean Ray* won't be out until late in the afternoon, but *Midland* says she wants to come onto location now. We go looking for our markers in the dark, the beams of our searchlights staggering over the water like translucent stilts. There lies *Midland*, a monument twinkling on the horizon, a ruby high on the crown of the derrick, an emerald and a diamond right below. We're having trouble finding the flags. Whitecaps fool us, seagulls following the searchlights fool us. We find number three, but the rest is a mess. Rather than cut any more anchor lines, we'll go back to Exxon in the morning, recalibrate, and then come back and fill in the holes. *Midland* doesn't like it, but that is the way we're going to do it. John E. Chance says so. Al, smiling apologetically, says sometimes it gets a little tense like this, but it all sorts out. It's just that these people are used to Gulf weather.

Kurt says, "You're not supposed to hold up the rig. That's a cardinal sin." He calls *Alaskan Star* for permission to come back and visit. "This is where it gets to be fun," he says, "running the damn line between the plans and the weather."

Monday looks like the day. *Midland* is close in, the two stocky tugs out on their leashes. *Midland* is, in the language of her owners, Rowan International, Inc., a semi-submersible, two-pontoon, eight-column, stabilized offshore drilling platform. She rides now with her pontoons awash, each a little shy of three hundred feet long and forty feet wide. Her columns, or legs — the corner ones have a diameter of twenty-seven feet — support a two-story deck that has air-conditioned quarters for ninety-four men (there are few women on U.S. exploratory rigs), a mess for thirty, laundry, showers, and a six-man hospital. The deck covers a little less than an acre, but it looks smaller. That is because it is crowded with a deck house, machinery sheds, vans, pipe racks, the substructure for the 160-foot drilling derrick, a heliport, four double anchor winches, and assorted rescue craft. She is nowhere near the largest semi. That one, *Ocean Ranger,* is drilling up north off Newfoundland on Mobil's Hibernia oil discovery. But this morning, the tip of her derrick the length of a football field above the sea, *Midland* has her own magnificence. Her profile will shrink in a while when she ballasts her pontoons and settles until her keels ride a steady fifty feet below the surface, a proper platform for the drill.

None of that can happen without the anchors, and the tidal currents are playing with the markers again. Half the flags are under water, so *Millie Bruce* will have to stand by them to guide the anchor boats. After *Saratoga*'s misfortunes Mobil decided to run the primary anchors even further out, almost a mile.

We're ready, *Midland*'s ready, but we're still shy an anchor boat. Capt. Mack is muttering. "Shame we didn't start nine o'clock this mornin'. Shame the weather's bad. Shame we wasn't born with a silver spoon. Goddamhell," he hollers. "We deserve a break."

At this moment, the snippet of confused communication I told Halley the Icelander about begins on the radio. The lobsterman *Bronco* inquires of *Midland*'s intentions. When he sees the location is away from his traps, he signs off with that "Finest kind." It comes out "Foinest-koind."

Capt. Mack asks the Yankee interpreter what the lobsterman is saying.

"Good," I say.

"Coonasses won't understand that."

They don't. They get "pie in the sky" out of it. But that is not a hostile translation. There's something soothing about it on this ceaseless sea.

Jarl's skipper backs in under *Midland*'s crane, which hands him the pendant line from the number-four anchor. *Jarl* gingerly winches the great iron spearhead off its roost on the pontoon. He guns out to a marker and lets the anchor smash into the water and marks it with a can. *Ocean Ray,* late out of Davisville to begin with, joins in. Half an hour later, he has blown his winch. That means fifteen hours back to the beach, unknown hours for repair, fifteen hours return. *Jarl* will have to single it.

After supper, Al and I pack our gear and get into survival suits for the transfer to *Midland* to complete the positioning. Capt. Mack backs in as close as he can. The seas breaking on the pontoons look mean in the searchlights. The personnel basket drops out of the night. We scramble and jump for its ratlines. One jerk and the wind has us. We soar over the black roil and into the hands of the *Midland* crew. The basket flies out again after the surveyors, and Al and I stand on a steady deck — just a twitch now and then as the sea gives *Midland* a snide shoulder.

Massachusetts to Morocco didn't use to be much of a trip. Not, say, two hundred million years ago, just before the beginning of the Jurassic Age. The continents were grouped then. The theory of plate tectonics, or, more precisely, the theory of seafloor spreading, holds that the surface of the planet is a jigsaw of thin crustal plates, some with continents riding them like topping. Most lurch around at a centimeter or two a year, overriding their neighbors or being overridden by them, the up-thrust margins forming mountains and otherwise rear-ranging the furniture, the trapped margins diving deep to melt in the furnace of the asthenosphere to the tune of earthquakes and eruptions. And so, in the pre-Jurassic logjam, the continents of Europe and Africa and North America and South America lay together. Another ten or twenty million years and the earth split, the cracks reaching to the crucibles below. Magma, molten rock, rose in the fissures. The modern Atlantic (there are indi-cations of an older one) crept open.

Geologists at Woods Hole preserve an old, small joke, told at the expense of the rival Scripps Institution of Oceanography, perched on a scarp of the California coast. Our ocean is growing, they say. Theirs will be a pond soon, and if those guys don't watch out, they're going to disappear under the Aleutians. Woods Hole looks out on the continental margin of New England. "Continental margin" is a collective term for the rela-tively flat and, in this case, wide continental shelf, at the seaward edge of which the somewhat steeper continental slope descends to the gentler continental rise that leads from the foot of the slope to the abyssal floor of the ocean.

New England's is a passive margin, a trailing edge, a stern to Scripps' prow.

Since its earliest researches, Woods Hole has spent a good deal of its time examining the banks and basins of this boundary between continental and oceanic crust. Its geologists and geophysicists have gone far out to sea, to the ridge that runs, like a baseball seam, the length of the Atlantic and on into other waters, to see for themselves how the welling magma builds new flooring for the ocean. And they have occupied their stations over bottoms shelving toward Maine and Massachusetts, inferring the beginnings of what was, once upon a time, one with Africa. Some, the academics, have operated more or less on their own. Others have worked in the public sector, in the economic interests of the country.

David Folger and John Schlee are government scientists. They work in the Woods Hole office of the United States Geological Survey (USGS) in a low building, set among scrub oak, that used to house the Oceanographic's collection of cores taken from the bottom of the world ocean. Their group is responsible for informing the government and the public on nonrenewable resources of the outer continental shelf — that part of the margin seaward of state waters — from the Canadian border south to Florida.

Dave is no stereotypical Yankee. He is open, with a welcoming face. Old-line he is, though. His people were Nantucket whalers. One set a cousin, Ben Franklin, straight on a navigational point of some importance: when Franklin wanted to know why it was that some British packets were taking weeks longer than others in their passage to the colonies, Dave's forefather explained it was because they were breasting the Gulf Stream. He said he and the other whalers had tried to tell the British that, but "they were too wise to be counselled by simple American fishermen." Dave himself likes to get as close to seawater as he can, flitting around in small sailboats and swimming difficult distances — many times the six miles across Vineyard Sound and once more than double that, from Martha's Vineyard to Nantucket.

Dave has taken a close interest in what the oil companies have been willing to tell him about the formations they have penetrated in the Atlantic offshore. They have kept a lot to themselves, proprietary information that might help them in the competition. But Dave has learned something here and there.

In a lot of its work on the offshore frontier, USGS, like the oil companies, depends on echoes — reflections, really; bounces of sonic energy back from the leaves and layers of rocks under the seabed. Explosives were used originally to generate the energy, but they were hard to regulate. Occasionally they blew someone to bits. Fishermen complained of the damage done to their prey. Now, multimillion-dollar ships, really sea-kindly computers, run thousands of kilometers of seismic lines, using clusters of air bangers submerged close to the ship to create a complex of sound and trailing hydrophones on long cables to catch what comes back.

The simplest way to describe the process, Dave said, is in terms of density contrasts at various depths. "The problem is complicated because the sound and energy you propagate down there bounces around between different interfaces — different kinds of rocks, different kinds of conditions. For example, a lot of water in the sediment may change the acoustic characteristics. So can the presence of oil and — especially — gas. When you replace a fluid in the pores of the rocks with gas, that changes the whole medium. Also, salt has a different sound velocity from other sediments at equivalent depths. Salt is light. It tries to rise, just as if it were oil. If it is deep enough, it makes domes or diapirs, just like bubbles."

The air banger releases charges of air under high pressure just under the sea surface. (New devices using water instead of air are coming into use.) The bigger ones, producing more than one thousand cubic inches of air in an instant, create a boom with a low frequency. "That energy will penetrate a long way," Dave said. "It can go into the upper mantle, but we're not sure what we're getting at that depth because we don't know much about the velocities."

That is the trick. Different rocks and rock structures affect the velocity of the signal in different ways, and the trained geologist can use the difference to guess what lies under his keel. To refine those guesses, clusters of bangers are deployed, producing a suite of frequencies. The data are recorded, enhanced, massaged. The information moves pens across meters of paper to produce seismic records whose lines can be as dense as fabric. With these and other printouts, the interpretation teams — the Merlins of the game — make their divinations.

More often than not, the frontier fools the experts. Certain interesting breaks and bulges do show up on modern seismic records, but other more subtle and often more interesting phenomena do not. False readings are a plague. The oil companies put down a lot of money on Destin Dome off Florida, Dave said. "They have developed a seismic record that produces what they call bright spots. Bright spots mean gas, and Destin, I'm told, had bright spots. But it turns out that other things will give you those readings. Thin layers of volcanic rock. Certain kinds of fluid clay layers." No gas has been found at Destin, and no oil.

On Dave's wall was a map of the North Atlantic showing the seismic lines run under USGS supervision. They were widely spaced, twenty to forty kilometers apart, but close enough to give a regional view of the margin. The oil companies shoot their own lines, and those are apt to be much closer together in areas of interest. But in going about its work of gathering information for its employer, the Department of the Interior, about what resources may lie where under the federal sea, the Geological Survey often helps oil scientists sell prospects to their own companies. This happened, Dave said, on the Atlantic margin: "John Schlee and I gave our interpretation at a conference, and afterwards, two geologists from one of the majors came up and told us they had been trying for months to get their people excited about a basin out here. Our presentation, they said, was just the ticket."

With enough excitement comes a COST well — COST for Continental Offshore Stratigraphic Test. A consortium of interested companies drills it under USGS direction. They do not want to find oil. In fact, a find would be embarrassing. The government is not supposed to be involved in that sort of thing; in this game, you have to be private to play. And a find would give everyone a devil of a time figuring out who owned what. "The purpose of COST," Dave said, "is to calibrate all your seismic information. You immediately get a velocity profile down the well. Then you simply take that profile and say, 'Here is a reflector, and it's at this depth, and *that's* what caused this squiggle on my record.' Then you can extrapolate. You can't go too far, because the character of the rocks and thus the velocity may change. But if the beds are relatively coherent and relatively homogeneous over a large area, you can do pretty well. For example, there are limestones exposed as far west as the Michigan basin that have physical characteristics almost identical to the limestone deposited up against the west side of the Adirondacks. So you can begin to get a handle on things."

Two COST wells were drilled in the midseventies on Georges Bank. They were sunk within fifty miles of each other near the base of the great thumb. The deeper one went more than four miles down. The wells provided enough of a handle on things so that USGS could get on with its mandated chore of assessing oil and gas potentials. "The method," Dave said, "is to look at the general geology — the kinds of rocks we think are there — and then make comparisons with basins that have been studied and drilled."

Folger and Schlee, solo and in concert, have used the COST data, together with information collected over two decades of records, to put together their version of the geological history of Georges Bank. They subscribe to the theory that somewhere around the late Triassic, say, around two hundred million years ago, rifting began along the line that would eventually mark the separation of Africa and North America. The brittle crust of the continent fractured and broke, and the molten igneous rock drove up through the cracks. Lakes probably filled the hollows along the separation zone, much as they have in the Rift Valley of East Africa.

"As the separation widened," Dave said, "Georges Bank lay at the bottom of a very narrow sea like the Red Sea. The climate then, a hundred and ninety million years ago, was obviously arid and probably hot — what we call a sabkha environment. There was a narrow opening to the ocean, and seawater kept pouring in and evaporating."

Fracturing continued, and, as the sea widened, the hot rim of the new American continent cooled and subsided. An irregular depression or basin formed where Georges Bank now rises, its deepest point under the south central part of the Bank. The broken blocks of the basin were buried under kilometers of muds, sands, and organic materials from the waters above and from the new rivers. At intervals, volcanic rock forced its way into the sediments, distorting and faulting them.

In time, what had been a restricted marine shelf on which sand, salt beds, and limestones were being deposited became an open shelf — again, something like the coastal shelves of the Red Sea today. Coral reefs began to build along the seaward edge of the basin, and the polyps kept pace with its subsidence by building their great carbonate banks upward toward the light.

Sediments filled the basin itself to a depth of about eight kilometers. Then, said Dave, "the reef finally got buried by the terrigenous sediments in the lower part of the Cretaceous, about one hundred and twenty million years ago. Some patch reefs persisted, but eventually the whole outer edge drowned under two or three kilometers of muds and silts and sands. That material, a lot of it, went right by the shelf break and accumulated down on the slope, far down on the continental rise."

On its surface, Georges started out as a cuesta, rising gently from north to south and then falling off sharply. Glaciers of the Pleistocene locked up so much water that the sea level fell almost 400 feet, according to some estimates; Dave thinks 240 is probably a safer figure. A hundred and fifty centuries ago, the last glacier covered all of New England and part of what is now Georges. Moraines bulldozed by the glacier added flesh to the thumb, and rivers of meltwater cut out at least some of the channels and canyons. "Georges would have been connected to Cape Cod then," Dave said, "but the Northeast Channel would still have been covered with water." He looked over his shoulder at the great blue chart of his turf and gave his easy laugh. "They found mastodon teeth out there, you know," he said. "And if the mastodon were there, I guess we were too."

John Schlee often acts as point man for USGS in publicly interpreting the basins of the Atlantic margin, particularly Georges Bank. His presentation is meticulously scientific, but even while he declaims upon strikes and dips and horsts and grabens, the intensity of the man comes through. Sloppy questions irritate him. They seem to jar his mental clockwork. Giving a paper or going through the figures of a Scottish country dance down in the Woods Hole community hall by the drawbridge, John never stops ticking.

One day in February of 1980, a few months after Frank Basile had his good day in Providence, Schlee went up to Boston's famed hearing house, Faneuil Hall. (For all I know, he may have gone by bicycle. He uses a bike as others use a car, butting along in his steady pace. For fun, he and his wife, the writer Susan Schlee, once biked across Iowa.) John had been asked to appear before members of the Senate Committee on Energy and Natural Resources studying the impact of oil and gas development on Georges Bank. Two senators handled the proceedings. Paul E. Tsongas, Democrat, the junior senator from Massachusetts, was on record as being concerned that the federal government's accelerated leasing program wasn't paying enough attention to the fisheries. Lowell P. Weicker, Republican, the senior senator from Connecticut, had made a name for himself as a fighter for a national ocean policy; the summer before, he had taken testimony about the huge blowout of the IXTOC well off Mexico and was worried about what a similar spill could do in American waters.

Both men wanted Schlee to tell them what his organization thought lies under Georges Bank. The United States Geological Survey is required by law to think such thoughts, the idea being that it ill behooves the Secretary of the Interior to lease public lands without first having at least some idea of what he is selling.

There are, John told the senators, certain prerequisites for the presence of oil and gas in commercial quantities. There must be source rocks — organic-rich shales, for example — buried at a depth sufficient to encourage the generation of hydrocarbons. Second, there must be porous rocks to serve as a reservoir, and they must be in a position to receive the oil from the source beds. Third, rock impenetrable by oil must lie over the reservoir beds as a seal to keep the hydrocarbons from seeping upward and escaping. Fourth, structural or stratigraphic traps are necessary to trap the hydrocarbons. Otherwise, they will keep migrating and dissipating. The final item, John explained, is timing; traps and seals and reservoirs must be in place *before* the oil and gas migrate from their source beds.

Georges contains rocks porous enough to function as reservoirs for oil, John said. The old reef at the seaward edge of the basin, if it is like other reefs, should contain potential reservoir rocks interleaved with tighter or less porous formations. Red sandstones and shales lie in the basin deeps and some of the limestones laid down in shallow water probably have turned into more porous dolomite. There is no shortage of traps either. Some were formed by sediments draped over the great blocks of rock thrust up from the bottom during faulting. Salt from the early seas ballooned up in spots to create other inverse cups. Then there is the reef itself.

The big question is source rocks. "A decade of drilling in the Scotian margin, northeast of Georges Bank, has shown that source rocks in the oldest and deepest part of the basin are lacking except in a few thin coal beds," John said. Seaward of the basin, out where dark, organic-rich clays formed in bottom waters low in oxygen, things might turn out to be oilier — if the clays were buried deep enough to get hot enough and if there were enough pathways for the oil to work its way shoreward to the reservoir reef.

Georges being rank frontier, nothing could be solid about resource estimates there. "The Canadians have been looking at their offshore a lot longer than we have. The first drill holes on the U.S. East Coast were in the Baltimore Canyon two and a half years ago, whereas the Canadians started in nineteen sixty-six, and it hasn't been until last year that they had really major commercial finds of oil and gas." Mobil Canada had the major hand in both, the first the Venture gas well off Nova Scotia, and the second the big Hibernia find off Newfoundland. "So," said John, "after fourteen years, they have begun to find some commercial oil and gas in their off-shore."

What USGS had come up with, and it is plain from talks with assessment people there that they were only doing it because the law required them to, was a high-low game. There was a 95 percent probability of finding nothing under Georges, John said, and a 5 percent probability of finding at least 2.5 billion barrels of oil (a barrel contains forty-two gallons) and at least 13.2 trillion cubic feet of gas. In between, take your pick. Fifty-fifty? About a billion barrels of oil.

"To develop offshore takes a lot bigger find of oil and gas than onshore," John testified. "The Gulf Coast — I think the figure I hear mentioned is twenty-five million barrels of oil to warrant construction of the platforms and so forth connected with development."

"But that's a function of the world price of oil," Senator Tsongas said.

"Right," said John, "and it's going up."

"I noticed," said Tsongas.

John Hunt the geochemist is as cautious as Schlee the geologist. Hunt's research and reading has indicated to him that "the Atlantic margin is not another North Sea and never will be." Hunt, in his midsixties, has a sandy, Celtic look. He is a senior scientist at the Oceanographic and spent sixteen years working in the oil industry.

Around Woods Hole, Hunt is to the preparation of oil and gas what Julia Child is to the preparation of good food. He has been known to make use of the same terminology in describing hydrocarbon generation for the layman. Temperatures in a sedimentary basin increase as you drill deeper, he wrote in the Oceanographic's magazine *Oceanus*. The increase is variable but it runs around 3 degrees centigrade per hundred meters or 1.7 degrees Fahrenheit per hundred feet. "This heat cooks the organic material in the same way that you would cook a roast in the oven. If the roast is cooked for the proper time and at the proper temperature, it will turn brown and yield an oily liquid, depending on its fat content. Likewise, the organic matter of rocks turns brown and yields oil. . . . If you let the roast heat for too long a time at too high a temperature, it will turn black and give off smoke (gases). Likewise, organic material which is buried very deep for too long a time will turn black and yield only gas (methane). In contrast, if the organic material is buried to a shallow depth and not heated, it will sit for millions of years without making any oil. Thus many rich oil shales are simply source rocks of petroleum which have never been heated high enough to yield oil naturally.''

Petroleum hydrocarbons are renewable, if you are willing to wait millions of years for renewal. Algae and other organisms die and drift in patchy but unending rain to the seafloor sediments. Rivers carry detritus from banks and marshes out to the marine bottoms. Some hydrocarbons occur naturally in the fresh organic matter that is seaborne — waxes in seeds and plants from land, liquids in marine organisms. But by far the greatest quantity of hydrocarbons is generated deep in the sediments during thermal alteration.

The process is profligate. Worldwide, according to Hunt, the average shale contains about 1 percent organic matter. He calculates that about a tenth of that is converted by heat and time into bitumen, of which a tenth migrates out of the source rocks as oil. Half that oil is collected in traps — half of 1 percent of the original organic matter in the source rock. Only 30 to 40 percent of the oil found is recoverable given today's economic and technological conditions, though 80 percent of the gas can be extracted.

Prospecting for oil has little to do with the average. Quantities and qualities of organic deposits vary immensely within and among basins. Marine organisms are apt to yield oil. Land-derived materials are better gas pro-

ducers. The end products of the cooking process are gas
and graphite, but a range of alterations is possible along
the way. Optimal cooking temperatures in centigrade
are in the 50-to-150-degree range for oil and somewhat
higher for gas — 50 to 200 degrees. These spreads are
often referred to as the oil and gas windows.

"We know," Hunt wrote in *Oceanus*, "that oil will be
found in young, hot basins or in old, cold basins as well as
those intermediate in temperatures and age. It will not be
found in young, cold basins, since the temperature and
time have not been sufficient to generate oil. Also, it will
not be found in old, hot basins because the oil would have
been converted to gas." Hydrocarbons can occur at almost any depth in sedimentary rock (many shallow deposits are the result of biological rather than thermal
processes), but Hunt sees a sedimentary floor — where
the temperatures are in the range of 110 to 160 degrees
centigrade — below which is the realm of gas.

"The stuff that exists in the mature zones off the Atlantic coast is more terrestrial-type material," Hunt told
me in his long, thin office at the Oceanographic. "That
tends to yield gas and condensate. Condensate is a very
light gasoline that comes out of the earth as a gas and can
be condensed to liquid at the surface." Hunt pulled down
some reports from his shelves. "Look here. See how the
anoxic conditions were more prevalent along the European and African coasts than they were along the U.S.
East Coast at the time the organic materials were being
laid down. So preservation was greater over there. And
look at this. That second COST well on Georges bottomed
in Jurassic rock, smack in the middle of land-derived organic matter — stuff like coal that makes gas. The Hibernia discovery showed an algal-type organic matter,
oil-making." Hunt said he had never seen samples of
those rocks. "The only one who has them is Mobil, and
they haven't published on them as far as I know."

In the North Sea, Hunt said, the source rock, the

Jurassic Kimmeridgian shale, goes all the way from Norway down south of England. "What we have here are little patches of good oil source rocks, and we don't even know where they are. Mobil found one, and probably they are gambling that there may be others. But the wells drilled so far down here don't show any."

The hunting machine is a piece of pipe. It comes in various diameters and is usually about thirty feet long. Three lengths, or joints, to a stand. Stand screwed to stand to form drill string. Drill string lowered down through more pipe — sections of huge tubing called risers that together form a conduit safe from the sea, from just under the drilling floor to the wellhead on the ocean bottom. At the tip of the drill string is the bit, a steel fist turning on rollers set at the bias and fitted with teeth of tungsten or industrial diamonds, their length and composition chosen to fit the formations. At the top, the pipe is connected to the kelly, a long, hexagonal tube suspended from a grandfather of hooks that is in turn hung from a great-grandfather of sheaves that moves up or down in a warp of cables connected to the crown of the derrick. The kelly is held in the jaws of a rotary table. The table turns to the right, turning the kelly, the drill string, the bit.

Drilling mud moves under variable pressure down through the drill string, out through the nozzles in the bit, and back to the rig through the annulus, the space between the drill pipe and the wall of the well. Close to where the bit has been working, the wall is often bare rock, but at scheduled points during the drilling, still more pipe clanks down the hole — casing sections cemented to the wall to keep the formations intact and what may lie in them away from the hole until the time comes for testing. The drilling mud carries the cuttings from the bit up to the rig for examination and disposal. It lubricates the drill string and protects the uncased hole. Its weight, almost instantly adjustable by changing its composition, counteracts pressures from the formations that otherwise might send salt water or gas or oil or all three thundering up the bore hell-bent for blowout.

This paradise for plumbers is among the most contrary of our exploitative procedures. Learning what to expect takes decades. Many who try their hand at drilling never can, literally, get the hang of it, the quarter-million-pound hang of a steel needle inches wide and miles long, turning and belling and shrieking and working on thirty-seven new ways a day to do anything but drill. Some do, usually by working their way up through the years, with time out for training courses. You start, usually, as a roustabout or gofer, then you graduate to roughneck or pipe tender. Then to driller, who drives the probe, tool pusher, who bosses the shift. Then maybe to rig superintendent, if you work for the rig owners, or to drilling superintendent, if you work for an oil company. A few of the more administratively minded go back to the beach and on up through management to the corner offices high up in Houston or New York.

Mobil's drilling super aboard *Rowan Midland* is B. D. Garvin. He started sixteen years ago on land rigs. He quit a lot, he says, and got run off a lot. He is tall, close to six-three. He is a walking *T* for Texas. His shoulders come off a football lineman. His waist is still discernible, though you can't see all his belt buckle all the time. His head is a little too big, even for that big body, and his hair is black and sparse and uncommonly long for a Texan's. His mouth takes up a lot of his round face. The fact that he laughs a lot is reassuring. Bob Garvin with a reef in his eyebrows can be unsettling.

It is ten days before Christmas of 1982. Dawn shows dirty through the drip of sleet outside the Airport Sheraton on the edge of Providence's Green Airport. Some oil people snooze in the low chairs of the lobby. I think I recognize one and go over. Yes, it's Gary Delaney from a month ago when Al Mitchell and I spent a couple of days on *Midland* during the final positioning. Gary has been on jack-up rigs off Africa — the rigs with retractable legs that lower to the bottom and permit the drilling platform to be ratcheted up above storm-wave heights. Now he's learning about floaters, semi-submersibles like *Midland*, working with Bob. The reason I didn't spot him at first is that he has shaved off his beard. He says he did it last night and now he can't find his chin.

Here comes Bob Garvin, mean-looking and pouchy. He and Delaney were putting in time yesterday aboard a workboat, learning how to deploy oil-spill cleanup booms and skimmers. "Don't like working with people I don't know," Bob says. "Keep thinking they're gonna drop something on my foot." He sees me and smiles under his pouches. "Say, did you know that when you're bald in the back, it's a sign you're a lover? And when you're bald in front, you're a thinker? But buddy, when you're like you and me, bald all over, you just think you're a lover."

By seven, we're out at a hangar off at the end of the field. The helicopter service for *Midland* and the other rigs out on Georges is British, with a long and good record of shuttling crews in the impossible weather of the North Sea. Bristow, it is called. Crew changes are on Thursdays for the drillers and Tuesdays for the service crews, the loggers and mudmen and the others whose vans nest on *Midland*'s decks, and for Mobil people. This is Tuesday. Young men in wild-West moustaches slouch and sleep. A big color TV hangs from the ceiling to lull the waiters. The news is on. An announcer in Providence says local opposition is growing to the plan announced by Ronald Reagan's Secretary of the Interior, James Watt, that could allow drilling within three miles of the New England coast. "There goes New England," says a young moustache, and grins.

"That'd be awful," Bob Garvin says, sticking a cigarette into his big mouth. "I'd hate to be that close to the beach and not be able to get to the sonofabitch." He goes on about how it used to rile him when he was drilling off Alaska and could see the glimmer of civilization in the night sky.

A Scottish Bristow man lets us know that it's freezing rain from 600 feet up. We may get out late, we may not get out at all. So we go to breakfast. Bob and Delaney and a man who is going to fix the rig's satellite communications system — direct dial from *Midland* to anyplace when it's working — have steak and eggs and hash browns. They look with kindly concern at my toast and coffee. When we get back, the Sikorsky 61N has been wheeled out of the hangar into the mist. We're going.

We grab for flotation suits. Bob gets a Medium he can't stand up in. He looks like an orange Neanderthal. The chopper pilot says he'll stop at Nantucket to top up the tanks, just in case the weather forces us to do some extra flying. We warm up, move away from the hangar, and lift off. Those with sensitive ears cover them with plastic muffs to cut the blast of the rotors. We come into Nantucket blind and taxi over to a fuel pump. Someone asks what the air valves are for on the survival suits, just below the knees. "That," someone else says, with authority, "is so if you fart you won't blow your legs off."

When the tanks are full, the British pilot rounds us up. "No point in staying here," he says. "They don't give political asylum in Nantucket." In another hour and a bit, we close on the rig. I can see it on the radar scope between the pilots. They ease down on the helipad. We get our gear, and the wet wind skids us over to the stairs down to the main deck. A pile of people are waiting on the stairs to take the chopper home after their two-week stint of twelve-hour daily shifts called tours, pronounced "towers." Twelve on, twelve off. I follow Bob down through the crowd and off along a corridor to the company man's office.

Richard Walker is there. He is Rowan's rig super, a globular, tight-skinned man in his early thirties. He is so solid I can't tell where the muscle ends and the fat begins. Maybe it doesn't. Richard is a connoisseur of coonass, and as soon as he spots me he takes up where he left off educating me a month ago.

"Sigh, Beee-oool," Richard says in deep Louisianan, "wut separates a coonass from a horse's ass?"

I've got him. Bob told me that one back at the airport. The Sabine — pronounced "Saybean" — River, dividing Cajun country and east Texas.

You can't starve in south Louisiana, Richard says. "They got mud bugs" — crawdads — "and coons. They'll eat anythin'. Ef it doan baht em fust, it's susceptible to bein' et." Never cold down there, he says, "never a day lak ee-yis."

"Let's knock that off," Garvin says in a low and serious voice, "till I can see what's going on here." We knock that off. Bob checks the reports left by Charlie Stevens, the Mobil super he has just relieved, and turns to the CRT screen displaying the drilling numbers he needs. The drill is 1,335 feet below the rotary table. Richard tells Bob we're in gumbo and turning harder. Sticky clay is gumbo to a Louisianan driller. "You can drill that crap," Bob says, "but you have to watch your drilling rate. Gumbo tends to hydrate or swell, and you have to have a special mud to prevent that. Otherwise, she'll cake and make mud rings in the annulus." You have to be careful which muds you use on Georges, though. The EPA will get after you if you reach for one not on the list or if you use a mud with oil in it and then discharge it.

Richard tells Bob that some hydraulic lines down on the wellhead broke this morning. He says the diving contractors sent their people down in the bell and they did a fine job patching things up. The divers are decompressing now.

Bob shows me the work plan for the well. It allows 145

days to get down to the targeted depth of 18,500 feet. We're not too far off schedule now, maybe a day. *Midland* spudded in by opening a thirty-six-inch hole, cementing in thirty-inch casing, and installing a guide base over the opening in the seafloor. She has drilled a twenty-six-inch hole down to 1,292 feet and lined it with twenty-inch casing topped with a connector pipe. Yesterday, the crew ran the BOP, the blowout preventer, a monstrous stack of rams and valves and lines that contains and diverts sudden kicks of pressure from below and that can, if necessary, crimp and seal the well pipe itself to pinch off a blowout. The BOP, after its inching descent along cables running to the guide base, now sits atop the connector pipe, constantly monitored by television and by fish.

The well will be an upside-down telescope. Bob says we'll drill ahead on a 17½-inch hole to a predetermined depth, then set 13⅜-inch casing, then drill a smaller hole and set smaller casing until the bore gets down to 8½ inches. "If we run anything after that, it'll be seven-inch liner," Bob says. Casing goes all the way to the top, usually, one diameter inside another. Liner hangs from up inside the casing above it.

Right now, Gary Delaney is out on the main deck with the Halliburton man to run a leak-off test. The Halliburton machine is a red jumble of mud tanks and pipes and dials. It forces mud under pressure down the drill string and into the bore. When the dials show a levelling off of pressure, that means the mud is leaking off into the formations. The test shows whether you can safely drill down to your next casing point without losing well control. Delaney says no problem. It took the equivalent of mud weighing more than fifteen pounds per gallon of fluid to leak off. *Midland* is using eleven pounds at present.

"Well, let's go visit," Bob says, unfolding, tucking his shirt in under a belt that has "Bobby Garvin" done in brown and white leather across the back. He catches me looking at it. "I'm the tailgate," he says, "the baby. Got five brothers and three sisters." He reaches for his hard hat. "They still call me Little Garvin back home." I wonder if they smile when they call him that.

Bob walks out onto the deck and heads for the moonpool, the opening below the drilling floor where the whitish riser passes down into the sea forty feet below. The wind and water are rising slowly and steadily, and the slip joint in the riser slides up and down. As it does, great tensioners and compensators, driven by compressed air, take up and slack off to keep things steady. "That joint will take twenty-seven feet either side of center before you spill riser all over the bottom," Bob says. "Before that happens, though, you better get ready to unlatch from the wellhead. If you don't, you'll end up pulling off location and fishing for that riser with a crane." He tells me about a bad storm in the North Sea when the heave got critical. "I wore me a path in the steel from my office to the moonpool. That well gave me an ulcer, just staying there eight hours staring at the slip joint."

We climb the steep and rain-greased stairs to the drilling floor. Wind walls provide some protection. So does a gale of hot air from a heater as big as a house on the main deck. But the cold rain falls through the angles of the derrick steepling above us. The drill floor is smaller than the one I remember seeing on a rig working the Baltimore Canyon — the *Zapata Ugland*, now drilling off the Maritimes. But the tools are familiar: the power tongs, suspended on wires, used to connect and disconnect pipe; the iron roughneck, which does the same thing faster. There is the mousehole, where you put pipe you want to have handy, and the rathole, for the kelly when it's not in use.

The driller has his place of business over by the tool pusher's office. Instrument consoles surround him and hand and foot controls to regulate the spin of the rotary table, the rate at which the hook bearing the weight of the drill string descends and, therefore, the rate of penetration by the bit. Dials tell him what weight is on the bit, what his hook load is, what the mud pressure is.

The driller is watching the roughnecks at their expert work, dropping metal slips into the jaws of the rotary to hold the drill pipe, throwing chains around the pipe or moving in with tongs to make connections. There are four of them right now, all young, all uniformed in the troglodytic colors of mud and wet. They are halfway through today's shift and two days away from a plane ride home to Louisiana or Texas, paid for by Rowan, and a fortnight of freedom.

Bob charges back down the stairs to the main deck to see what the service people are doing. *Midland* has been selected by the federal agencies monitoring drilling on Georges as a test site. A man from Battelle Laboratories takes samples from an oil-water separator for later analysis ashore to confirm compliance with the rig's discharge permit. His reports wend their way to the Environmental Protection Agency, which runs the permit program, and to what used to be the conservation division of the United States Geological Survey. Since the advent of James Watt, that outfit is now part of something called the Minerals Management Service.

A few steps away from the separator is the decompression unit, where the divers are returning to normal from their morning's deep and murky work with the broken lines. Around the corner is the van belonging to a service company called Exlog, where two men minister to banks of electronic devices. Some of the clicks and whirs end up as numbers on Bob's ghastly green screen. Bob scrunches down at a microscope. He peers and motions me over. In the circle of light under the instrument are grains of tan and gray: cuttings. "This is some streaky formation," Bob says. "First we got gumbo. Now we're into hard siltstone." He straightens up, and his hard hat hits a shelf support. As usual, he is too big for where he is.

As Bob leaves, he slams his fist into the side of the van. "Damn! If I could only see what is going on at the bottom of this hole."

We are out in the heavy rain again, going back to his office. It's about suppertime, and as I head for some dry clothes, I see Little Garvin sitting in his slicker, staring at his tiny screen. Depth: 1,350 feet. Rate of penetration: a flicker while the numbers change. Revolutions per minute: 55. Hook load: 189,000 pounds. Weight on bit: 22.9 pounds per square inch. Drilling fluid temperature: 72 degrees in, 75 degrees out. Flicker, number shifts, flicker.

The catered mess is bright and clean, and the boss cook is fat and freckled and sassy. Pizza is the popular offering. "Just like the kind mother used to make," says a roughneck fresh from the showers. "Po' boy," says the cook. "Probably accounts for why your daddy run off."

Bob and Delaney and a Mobil geologist are tearing through the pizza. The gale outside is turning into something worse, and *Midland* responds with a lumbering roll. The talk is guns and pickups so juiced up you can watch the gas needle move, and dodging bullets on deer hunts in Texas and Louisiana, and working overseas. This is only the second job Bob has had in the States in eight years. He misses the North Sea. It's partly the money. But it's also the opportunities. In Norway, Garvin supervised platforms — production monsters that can cost over a billion dollars apiece — and not just one middling semi.

And there's what you run into in New England. Bob says he met a nice-looking couple from Boston on his last flight up from Dallas. They asked him what he did, and he didn't tell them. "Hell, Bill," he says, waving his third tumbler of iced tea. "I believe in it. I've got the arguments. I just don't want the hassle."

The next morning is nasty. The sun makes a gold mist over a scudding sea. The wind is up over forty knots. Bob is on the radio to Davisville. "Mobil base, this is *Midland*. You hear the weather?"

"Yuh." Ed Hyman is Bob's boss ashore, an old hand at drilling, and a friend. Ed is the link with the Mobil Exploration and Production Services offices in Dallas. He tries to supply the rig with what it needs in pipe and drilling muds and cements and expertise, but between the weather and mechanical problems aboard the chartered workboats, he's got his problems. Right now, a boat loaded with casing Bob needs is down for repairs, and that means another delay. That means a bigger bill for Mobil, which is by now paying rig costs of somewhere between $85,000 and $125,000 a day. "Don't worry," Ed says. "We'll do something for you, right or wrong."

"I like it better right." Bob says *Midland* is heaving to twelve feet and heading higher as the storm closes in. "We're going to pick up and get ready to hang off," Bob tells Ed. That means he'll raise the drill string a ways and clamp everything left in the well with a hang-off tool held in the rams of the blowout preventer. Then he'll disconnect the riser so he can be in better shape to take whatever shoving around the storm has in store.

Bob grouses to himself, worrying about his priorities — making hole against taking care. "You get criticized for anything you do anyway," he blurts out to nobody in particular.

"Science and skill," Ed says, with a chuckle. "Science and skill will prevail over ignorance and superstition." Bob laughs, and they sign off, *Midland* and then Mobil base.

Bob puts on his slicker and heads for the drilling floor. Now that the decision to hang off is made, he feels better. Downtime is expensive, but fishing for gear is more so. Besides, Richard Walker has just told him there's a leak in a flange on the blowout preventer. Bob stops at the TV monitor and looks, and there it is, a swirl of dark showing like blood in the water against the whitish mass of the blowout-preventer stack. Drilling mud. "We'll have to yank the thing," Bob says. "And we'll have to do it before the mud cuts a groove in the flange."

Roustabouts, roughnecks, driller, ballast man, radio operator, crane operator, motorman, mechanic, electrician, tool pusher — just about everybody on the shift knows we're tripping up and hanging off, most by sensing the change in the rig's rhythm and voice. The iron roughneck grabs pipes above and below the threads and spins them loose. Then the driller raises the hook and the roughnecks manhandle the stand — ninety-three feet of solid weight — over to the side of the drilling floor. The derrick man, high and lonely, comes out on a monkey-board and guides the top end of the stand to its place in the racks. Drilling fluid is pumped into the hole as each fifth stand comes free. "That's government regulation," Bob says. "Otherwise you could get a space at the bottom of the hole that could let stuff come in. Then you could have yourself the beginnings of a blowout."

Now the stands are clanging like the devil's chimes. The rig rolls, and the derrick man loses his balance and hangs in the air by his safety belt before muscling himself back to his perch. A stand gets loose from the roughnecks and flies across the floor toward the driller, toward me. I fly into the tool pusher's office. The driller calmly lowers his hook and grounds the pipe. Bob nods at me and breaks out laughing. "You're learning," he says. "First thing you find out in the oil business: when you see a roughneck run, you run faster."

Something bangs below. The seas are running to twenty-five feet and the wind is over seventy knots. We've got the extra-tropical version of a hurricane. Bob worries that he waited too long to hang off. They've run the hang-off tool and set it in the BOP ram, and it doesn't want to release the pipe that lowered it. So Richard Walker takes over. He has a huge wad of snuff bulging out the front of his lower lip. He counts cadence for the roughnecks trying to break the pipe loose. Fat turns to fury. Richard screams. He dodges in and throws a chain around the pipe. It gives up. The driller starts to raise it, and Richard hollers again. The control hose for the block compensator system is caught. The roughnecks jump as Richard gives tongue. One vaults into the derrick and releases the hose.

"Now, you gotta respect that," says Bob and roars approval at Richard. The men are cleaning up and hosing off as we leave. My boots slide off the stairs and I make the journey down fluttering off the rail like a pennant. Bob waits at the bottom. "My gut is going to be bad for a week," he says.

Bob calls the beach to tell Ed we're hung off and ready to unlatch and move off location if we have to. Ed takes the message as if it were a request for some more Barite or other drilling staple. We go to lunch. Steak or macaroni and cheese and French fries, beans, and three kinds of dessert. "No hats, muscle shirts, shorts, greasy clothes or work shoes in dining area," a sign reads. We eat and listen to the storm outshouting Richard Walker.

We are taking a lot of spray across the deck now. The sea is coming to visit. Richard tells his people he wants a check on anchor lines. All hands use the buddy system and wear life vests — though it isn't immediately clear what good they would do if you went over the side and dropped into that frigid frenzy. If it gets much worse, Richard says, he wants life lines rigged. But he doesn't believe the waves will top fifty feet. "You don't get the fetch here you do in the North Sea," he says. Fetch is the distance over which waves have a chance to grow.

Those with no business on deck keep clear of it. The Mobil geologist gives me an introductory course in some of the logging techniques used to signal what is going on down-hole. Some of the logging is done while the drilling is in progress, some when the drill string is raised and special sensors are run. Technicians send electrical charges into the formations to test resistivity and conductivity and thus the porosity of the rock. Radioactivity sensors help identify clays and shales. When you have something interesting, you take a core sample and send it to Dallas. If Dallas gets excited, it will ask for testing of the interesting formation. Then you lower a gun tool, seal off the test section, and pull the trigger. The gun fires projectiles through the casing and into the rock so that whatever is in it can flow into the well and get circulated up to the rig for identification.

Richard comes in about nightfall to tell Bob we've lost the number-eight anchor. We are unlatching from the wellhead. The risers will be snugged up against the bottom of the drilling floor to give *Midland* more clearance. For the next hour, Richard is all over, checking back with Bob, who is checking with Ed at Mobil base. Two more anchor lines go. We're about twenty-five feet off location. "How long would it take to crank up Bristow?" Bob asks the beach.

"About an hour," Ed says softly. How could any chopper fly in this?

"Well, maybe you could let 'em know. If we go adrift. I want to get all the service people off. I just want what's necessary here."

"I'll call 'em," says Ed.

Richard lurches into the office, wet through, an enormous walking raindrop. He gets into a chair, for the first time in twelve hours, and leans back. He has lost his voice and speaks in squeaks. The storm, he says, is easing, and the remaining anchors look as if they'll hold. He has told his people that they are in a hurricane and that they aren't in the Gulf. There, he says, they shut the rig down and get the crew ashore before the main show. "They got the message they wan't nobody comin' for 'em out hyah." He puts his hands behind his head and squeaks at the ceiling. "Ah told 'em to remember wut Ah told 'em. Ahrn has no feelin'. It doan care how much of you is between it an' the wall when it hits you."

Next morning, Bob is talking to Mobil base. He says he's got a service boat out of service, the *Tampa VI*, a converted party boat with special attractions for the angler, like heated handrails. *Tampa* called in the middle of things last night to say that the seas had knocked out her port windows and she was drifting, already twenty miles away, lying in the trough to keep another wave from taking out the rest of her glass. Bob needs a workboat with an anchor-setting crew. "Next time," he tells Ed, "we're gonna have to get out of the hole earlier if we think there's a chance of breaking an anchor wire."

The weather report says a window will develop later in the day for a few hours before another low comes in. Bristow makes it out, and I make it back with them on the first helicopter shuttle. We fly over Great Point on Nantucket, foaming in the wash of the great storm. Down Nantucket and Vineyard sounds, and right over Woods Hole and New Bedford to Providence. When the rotors whine down and stop, we pile out. One roughneck has a black western hat with turquoise and silver on the band. The storm has dropped some snow on the airport, and he heads for a pile. Maybe he's got time, he says, to build a snowman before his flight to New Orleans.

Back in 1975, when it was first dawning on New England that its offshore was going to be drilled and that parts of its littoral might bear the consequences, a Nantucketer asked himself a question. "Over one hundred years ago," he said, "the discovery of oil at Titusville wrecked the whale-oil-based Nantucket economy. Is offshore oil exploration and production going to wreck the present economy and replace it with a transitory one that will ultimately destroy the island and the quality of life as we know it today?"

When rock oil was brought to the nation in 1859 by a former railroad conductor named Edwin L. Drake, one of the purposes his backers had in mind was to refine it into an illuminant that would supplement and supplant whale oil. Up until the day when Drake brought in his primitive well just outside the village of Titusville in northwestern Pennsylvania, oil had been regarded as a messy by-product of drilling for other things, like salt. One driller got rid of the stuff by selling it as "Kier's Petroleum or Rock Oil Celebrated for its Wonderful Curative Powers. A Natural Remedy Procured from a Well in Allegheny County, Pennsylvania, four hundred feet below the Earth's Surface." Petroleum didn't do in whaling by itself. Things had begun to slip in the fishery before Drake and his successors and new lamps from Europe put kerosene in American parlors. But the transition from the one oil to the other was painful for Nantucket and New Bedford, and they remember.

The rush Drake started may have fathered the modern oil business, but it was a latecomer in the history of petroleum. People knew about the stuff thousands of years ago. They talked of a gas seep in what is now Iraq, calling it "the father of sound, where the voice of the gods issueth from the rocks." The Chinese were drilling to three thousand feet by A.D. 1100, and there is evidence that they found and used oil for fuel. At the time of our revolution, there were hundreds of oil wells in Burma. What we did was to make petroleum a way of life, into an industry as American as manifest destiny.

It was, many forget, an industry of the American Northeast, and it grew up dealing with large numbers of small landholders. The Europeans, who had financed so much of our earlier development, weren't much interested in the kind of high-risk investment Drake had started. Geologist Edgar Wesley Owen, who wrote a history of the business called *Trek of the Oil Finders,* said that in providing venture capital, "the American temperament was . . . inclined to shoot the works. Almost overnight, the blacksmith, the shoemaker, and their neighbors took their life savings from under the mattress and got into the oil business. The markets and the bankers were far away, the oil men were here. They were in oil country, and their time had come." Oil historian Ruth Sheldon Knowles wrote that "oil is the greatest single source of wealth in America for individual fortunes. At the same time, exploring for it is the greatest source of business failure, a fact to which wildcatters deliberately blind themselves." Knowles' writing has just the blend of excitement and hyperbole that permeates the oil patch. She calls one book *The Greatest Gamblers* and dedicates it "to all the unsuccessful explorers who have drilled America's more than 730,000 dry holes and whose failures have guided others to the discovery of the abundance of oil and gas which made America a great industrialized nation."

These days, the public is apt to think of the industry as monolithic: Big Oil, Seven Sisters. But throughout its history, the independents have been the oil hounds. They have often had help from the majors, true, but the small operators were the ones who set up shop under names like Hoppy Toad or Only Oil and went hunting. They used "creekology" and doodlebugs — dowsers. They relied religiously on a certain lady. "Finding oil is all luck," said H. L. Hunt, who won some of his best prospects in poker games. They failed by the tens of thousands and hit by the hundreds. Charles L. Woods was out in California when the century began. His nickname, according to Knowles, was Dry-Hole Charlie, until the day in March 1910 when "the derrick disappeared in a crater cut by a column of oil twenty feet in diameter. 'My God!' Charlie whooped. 'We've cut an artery down there.' " And so, on and on. Dad Joiner, the hapless one who stumbled on the east Texas field, the biggest in the lower Forty-eight. Mike Benedum, Joe Trees, Tom Slick. So many competing for so much that when a dry and eminently reasonable man called John D. Rockefeller came to review the nature of his corporate contribution to the industry, he pronounced it a bulwark against the excess of pluralism. Standard Oil, he said, was an "angel of mercy," picking up the pieces of businesses ruined by the depression of the 1870s (and by trying to buck Rockefeller's hold on 80 percent of American refining capacity and 90 percent of its pipelines). "The day of combination is here to stay," Rockefeller said. "Individualism is gone, never to return."

He was wrong. The days of combination come and go, along with individualism and just about everything else caught up in the cycles of the oil business. The Standard Oil Trust went under, in 1911, broken up by the Supreme Court into thirty-eight pieces. Today, some of the outgrowths of those pieces operate with more money than do some sovereign nations.

Rockefeller started on his way to the top believing in kerosene. Gasoline was just a waste product. But oil as fuel took hold quickly when it did take, when automobile ownership became an American right and when American industry converted its power source from coal. Americans went to the movies to see a gusher raining on the uplifted, laughing face of Clark Gable. So much oil was coming out of the ground that some got to thinking the country might run out of it. There was an oil-shortage scare before and after the First World War and a lot of talk of conservation. But these were as spasmodic as our recent scares appear to have been. "The oil industry has always been criticized in boom times for wasteful practices," Edgar Owen wrote. The criticisms were "as inherent in the legal doctrine that petroleum is a fugitive substance subject to the law of capture as they were expressive of the aggressiveness or cupidity of the captor."

Like the giant cod on the Atlantic banks, the giant oil fields don't turn up the way they used to. Giants are capable of yielding at least one hundred million barrels of oil and one trillion cubic feet of gas (except in the Middle East, North Africa, and Asiatic Russia, where the minimum is five hundred million barrels or three trillion cubic feet). According to Owen, some 328 giants were discovered in this country up to the midseventies, 90 percent of them prior to the onset of sophisticated seismic technology in the midfifties. Some of the most famous of them were stratigraphic rather than structural in nature, created by subtle changes in the oil-bearing strata that even today are hard to find, rather than by heroic faulting or folding. Giants are still found. But only 33 of the 288 discovered during the period 1968–78 were in the United States or Canada. An industry expert surveying discovery rates since the beginning of the century says that there have been dips right along — during the Depression, World War Two, the period of excess capacity during the fifties. But starting in the early sixties, there has been a long-term decline in the rates, even though exploration activity was — until the recent recession — at an all-time high. One reason for that, the expert says, is the decrease in the size of the fields that are discovered.

In the oil patch, they claim that big fields make big companies. The corollary is, the bigger the company, the bigger its need for big fields. It isn't enough any more to shine at refining and distribution, the downstream end of oil. You have to own the headwaters at the wellhead. The need to hit big, and the flattening returns on discoveries, turned the big oil companies toward the last frontier, the land under the sea, the margins that together account for a third of the continental land mass.

Going to sea for oil, as opposed to drilling for it in the shallows, is still a youthful enterprise. Risks generally are higher than they are on land. Costs are five to ten times higher, more if you have to worry about polar ice or other environmental extremes. The offshore play has made it definite: Houston has beaten out Reno as the gambling capital of a gambling country.

Large fields have been found out on the shelves, in the gulfs of Persia and Mexico, off Malaysia, Nigeria, Australia, Angola, Indonesia, New Zealand. The Russians worked the Caspian Sea and began looking around in Arctic waters. The North Sea blossomed like a black rose with the discovery of Ekofisk. Brazil, the Philippines, Spain, eastern Canada hit big. Alaska, with the huge coastal Prudhoe Bay find producing, shows promise offshore. So does China. But for all that, the oil industry in recent years felt a bit unwanted. The great gamblers met the great conservationists at Santa Barbara in 1969, and that battle became a war. Then came the revolutionary rise in oil prices in the midseventies and the fight over windfall profits. Jimmy Carter paused in his call to the energy barricades to blast the industry and its leaders. Mobil, he said, was "irresponsible."

It was different in the summer of 1981, just a couple of months before *Midland* squatted down on Georges Bank, when a number of the best oil finders in the country took the University of Delaware up on its invitation to come talk about the future of their business offshore. Ronald Reagan was in the White House. James Watt was remodelling the Department of the Interior. Optimism, the gambler's drink, flowed and bubbled. Look, the boosters said, offshore wells are already accounting for 9 percent of domestic oil and 22 percent of domestic gas production. If we play things right, we should be able to double those figures, easy. Look, they said, here's an estimate that about a quarter of the world's oil reserves are waiting to be found offshore. So what if we've used up half the oil there is. That means there's another half to go after. Senator Lowell Weicker, courageous as ever, visited the lion's den with talk about what oil spills might do to Georges Bank and its fisheries. The audience invited him to go down with some divers and see how the fish love to hang around their rigs.

Ruth Sheldon Knowles took a dark view of the country's hostility toward oil in the pre-Reagan days. She called it petrophobia. "The American eagle seems to have folded its wings," she wrote. "Still, what was true of the past can be true of the future. Oil and gas are not so much a prisoner of the earth's rocks as of the laws men make. If the eagle has the strength to break the bureaucratic bonds that bind his pinions, it [sic] can fly again." The keynote speaker at the Delaware conference, an extraordinarily successful oil consultant with the intercontinental name of Michel Halbouty, picked up that theme and took off with it. "There is no limit to what technology can accomplish," he said. And: "The geologist is attuned to the earth, so I don't want anyone to tell me I'm not an environmentalist, because I am; but I'm not a fanatic, a radical." And: "Whatever is best for the people will be determined by the industry and determined by the market." But what drew yells and ululations from the ardent was, "Ladies and gentlemen, petroleum is still king." If what the environmental movement was suffering from was petrophobia, it would be fair to say that what Mr. Halbouty was stirring up was petriotism.

Going to see Mobil is like taking a commercial flight: first, you put your carry-on baggage through an electronic scanner. Mobil is not the only cautious corporation in New York, but it has more reasons than most to be cautious. Several years ago, its headquarters — just across Forty-second Street from the old Chrysler Building — was bombed by Puerto Rican terrorists, and a man was killed in the blast. Once past the scanner, you are shunted to one of many receptionists, who checks your appointment and issues you an identification badge. With that passport, you are free to take the elevators to the home floors of the third-largest industrial company in the United States.

That is how Mobil describes itself. In addition to the Mobil Oil Corporation, it owns Montgomery Ward, the retail chain, and a printing company. It dabbles in real estate here and there. Counting all its conglomerations, Mobil operates in more than a hundred countries and employs over two hundred thousand people.

More conglomeration appeared to be in the works in December of 1981 when I first visited the company. The papers were full of Mobil and mergers. Mobil was after Marathon, its sixth attempt to pick up domestic reserves through acquisition. Marathon was fighting back. Its chief executive officer said that "everybody in the U.S. who suffered through the energy shortage has got to ask the Administration and Congress whether this kind of acquisition is going to increase the nation's reserves by a single barrel. . . . How much oil might Mobil find with the five billion dollars they're offering for Marathon? If they had taken the money they spent for Marcor [owner of Montgomery Ward] and used it in their oil exploration program, they might have found enough reserves on their own that they wouldn't need ours now."

Fortune magazine had come out with a piece on "Big Oil's Biggest Maverick." The article said that mergering in the oil business, be it successful or not (Marathon later would escape into the arms of U.S. Steel), is often a response to a lack of crude. Mobil's petroleum reserves in the U.S. have been shrinking, not as much as Exxon's and Texaco's, but still enough to warrant action. "At our size," Mobil's chairman, Rawleigh Warner, Jr., told the magazine, "we can't consistently replace production through discoveries alone. That's why you see these [takeover] attempts."

Mobil, according to *Fortune,* made a mistake about ten

years ago while trying to cut its dependence on risky reserves overseas. It decided to go elephant hunting — looking for fields big enough to slake its big thirst — and it did so mostly in the North American offshore rather than balancing exploration between land and sea. The company did well in places — oil at Hibernia and gas at Mobile Bay — although it did not rank with the most successful of wet wildcatters like Amoco and Shell. But it did not devote the attention it could have to the Overthrust Belt in the Rockies and other terrestrial projects that have proved out well. It has revised its strategy, but it has had to pay a lot more for onshore properties than it would have if it had started out earlier.

What emerged from articles like this and from interviews with those whose business it is to figure out what Mobil's business is all about was a picture of opposites. Muscle and urbanity. Hostile takeovers and the sponsorship of some of the best television programming available. The style was fascinating, nowhere more so than in its public affairs, run by a former labor lawyer named Herbert Schmertz. Mobil's famous op-ed ads, usually checked by the top management before appearing in the *New York Times* and other big papers, promoted a Gentleman Jim Corbett image, a slugger with taste. *Fortune* quoted Schmertz as saying, "There's almost a Pavlovian reaction to everything we do." The article didn't indicate whether Schmertz was bragging or complaining.

Bob Graves was the man I was going to see. He was a vice-president of Mobil Oil Corporation, in charge of the hunting done by the company's exploration and producing division. He was a large man with slightly hooded eyes who looked right for his job, sitting in his shirtsleeves, heavy but balanced. He had a reputation for memory, for knowing what is important to know about the 43,893 oil and gas wells in this country and the 8,000 overseas in which Mobil has an interest. He is Canadian by birth and retains the flatlands accent of his native Alberta.

Bob talked at first about the seismic boat Mobil had shooting lines out on Georges Bank. The data it produced showed a feature that did not appear on the group shoots — the work subscribed to by several companies. Mobil prides itself on the sophistication of its seismic interpreters down in Dallas, and the interpreters decided that what they were looking at in the proprietary numbers was a deep and promising reef fifteen thousand feet below the sea bottom, some of it under Block 312, *Rowan Midland*'s turf.

"Evaluating a sale," Bob said, "involves coming up with an estimate of the speculative reserves that might exist under a tract. We place a risk factor on those speculative reserves and then assign a value of so many dollars a barrel consistent with rates of return — the profit ratios we have to achieve in that high-risk environment. And once having arrived at those bid values, we then go to our executive committee and obtain approval to bid at those levels. Then we begin meetings with our partners." Bob leaned back from his big desk and looked out across high Manhattan. "It's a very subjective exercise. Every company goes through it somewhat differently. And within Mobil, if you were to put three different groups of geologists and geophysicists through it, they'd come up with different numbers.

"Mobil might look at that prospect on Block Three-twelve and say: 'Okay, we believe there's a reef feature there. It has a total potential thickness of a thousand feet, in round numbers.' And then we would look at reef prospects around the world, hopefully drawing the closest analogy that we can to that type of prospect. And then we'd say: 'Okay, based on all our information, if that reef is productive, the pay thickness, based on worldwide averages, will be two hundred and fifty feet.' So, working with that and the size of the feature shown by our geo-physics, we would calculate the reserve that could be present under that tract, what we call the speculative reserve. Now comes the risk factor. Again worldwide, the chances are only one in ten that a prospect contains any production at all. So we apply the risk factor, ten percent of the speculative reserve. We might then apply current values of hydrocarbons in the ground to that risk number."

Judging from comments I have heard elsewhere in the industry, Mobil might also do some price forecasting to establish what the value of the hydrocarbons (Graves was inclined to think they were mostly gas) will be eight or ten years down the line when production would begin.

"Our experience tells us," Bob said, "that if we drill prospects, nine of them are going to be dry and one of them is going to be successful. The hope is that in the one you hit you find not only the risk reserve but the speculative reserve. If you do, then you win out on the economics, providing you have the wherewithal to stay in the game."

We were nearing a topic that stirs up the majors considerably. "The economics of the individual field are not in any way a true reflection of the oil business," Bob said. "Yet individual field economics are what gets the industry into trouble all over the world. The problem is you find a helluva nice oil or gas field someplace, and the government looks at the economics of that particular venture and says, 'My God! There's no way you should be making that kind of money.' What they forget is you've got to make that kind of money or else you're out of business."

The Atlantic, Bob said, has had a few successes like Hibernia and the Venture gas find off Nova Scotia. But the dry holes, the dusters! Mobil drilled on a top prospect in the Baltimore Canyon, a structure called the Stone

Dome. It lost more than $110 million there. "And Georges Bank is a big gamble, there's no question about it. Exxon's yelling about a dry hole and they're drilling their second well. We're drilling our first. Boy, that's rank wildcatting at its best. We got some two hundred and forty million dollars riding on the outcome of that thing, and our chance, quite frankly, of finding anything —" He stops himself. "A funny thing about risk. We always go up there and tell the executive committee, 'This prospect has a fifteen percent chance of success,' and they chew on that for a bit. Really, what we should be saying is that this prospect has an eighty-five percent chance of failure. It's exactly the same number, but it has totally different implications."

The same play of light and shadow operates in a partnership of companies working a frontier like Georges, participants in a mating dance that would make a whooping crane look like a wallflower. The partners sit around a table. They talk technical about the blocks in which they're interested — the geology and geophysics, who shot what lines on what prospect, but they cannot divulge what their own reserve estimates are. The bidding begins, betting that will eventually create the pot the partners will take to the sale. The process is shaped by law and its records are open to audit by the Bureau of Land Management, but the participants run their own show. Bidding is designed to move in one direction: up. Partners can increase their bid or call for further technical discussion or reduce the percentage of the action they're willing to handle on a particular tract. Or they can walk out of the room. They cannot, as it is called, chill the bid. "At no point," Bob said, "can you say, 'Hell! Why, you guys are crazy. There's no way that tract is worth that much, and here's why.' "

The partners bid along, the turn going around the table, while the numbers get closer to the actual value they have worked up, each company on its own, for the

tract. "If the other partners keep on well beyond our evaluation," Bob said, "Mobil would probably ask for time out
so that the guys who are bullish on the tract can get up
and explain the nature of their bullishness. If they come
up with something we haven't recognized in our evaluation, what we do is call our people and say, 'Hey! These
guys have three more seismic lines than we did, and they
have better control than we do,' and request approval to
bid at a higher level."

It wasn't Mobil that went back to headquarters for
more money for Lease Sale Forty-two. It was Mobil that
got up and explained the nature of its bullishness, particularly about a group of blocks down near the main canyons of Georges' southern flank. "Our string of pearls,"
was the way John Goff described them. Goff handled the
negotiations for the tract acquisitions Graves and his
people were interested in. Trained as a lawyer, he was responsible for bringing together a workable team of partners and operating with their representatives to develop
the financial and legal framework for the joint bids that
emerged.

"I know," Goff told me, "that many companies had decided that the magnitude of the bids, the difficulty of the
work, the risks of working in an uncharted area would
require that they bid with others. As a consequence,
when we finally got to the point where we could move to
find partners, a number of companies we would have
liked to have bid with simply weren't available."

What Mobil liked in a partner, Goff said, was someone
"who has done work before the sale, who can make a
genuine contribution toward the refinement of the geology that we see there — jointly see there — because
that's the base of the whole bidding process." Mobil was
also looking for financial capacity in bidding, exploration,
and, with luck, development of the field. "There are lots
of companies with money. Not as much as we have,

maybe, but enough. There are fewer companies with the technical competence we are interested in."

Mobil signed with Union and Amerada Hess, both of which it had worked with in past sales. They agreed the interest would be 40 percent Mobil, 40 percent Union, and 20 Amerada. "But it soon appeared, after joint meetings started, that Union simply wasn't going to be able to agree to the bids we wanted to make on some tracts, and that meant they were going to have to drop out on those tracts." It was up to Mobil and Amerada to take up the slack there or find another partner. Soon. "I don't know exactly," Goff said, "but I think we were within two weeks of the bid date by then." It turned out that new money was needed, and Goff said he spent four or five days on the phone trying to find it among the smaller companies — "still big as hell, but smaller — large independents, we would call them."

Finally, Goff called Philip Oxley, president of Tenneco Oil Exploration and Production, down in Houston. Oxley probably realized at that late date that the bid group had fallen apart on the tracts he was being asked to come in on. "A man with that experience," Goff said, "has been through plenty of partners' meetings when somebody starts thinking, 'I'm not so fond of the geology and I don't even want to go to the sale.' Or, 'I didn't realize the kind of development problems that exist, and I'm scared off.' Things like that. But you can't talk to the newcomer about what has happened, and you can't indicate to him what levels of bidding have been going on, because that would be chilling the bid."

Oxley must have known Mobil's reputation for betting high on offshore ventures. But Tenneco said yes. "They were coming in kind of in the middle of things, and their assessment of the sale differed substantially from ours," Goff said. "Right up to the moment they saw the data produced by our seismic ship. They were just as struck by it as we had been. The amount of money they had authority to spend wasn't nearly sufficient to cover what we had in mind because of the potential of those reefs. So the meeting was shut down, and Mr. Oxley went back to headquarters and came back with a bigger pot. Union stepped in on some blocks, and we found another partner, Transco." On Block 312, Mobil put up 54 percent, Amerada Hess 25, Tenneco 12, and Transco 9.

"Mobil was viewed by industry as having bombed the sale," Goff said. "If I recall, we won every block we really wanted. Now, a lot of people wondered what the devil we were doing bidding those prices for tracts that attracted some interest but much lower bids. I remember that after the sale, Shell and their group approached us about buying into a couple of our blocks or exchanging interests. It didn't come to anything, for the usual reason: we placed different values on what we had and just couldn't get together on it. But I believe Shell had a suspicion at the time that we were seeing something that industry didn't see."

Goff had to go to a meeting. "I can't cite you any specific cases," he said as he got up, "but my experience tells me that time and time again we have seen something that industry hasn't seen. Maybe we didn't see it correctly." He opened the door for me. "And there have been many cases when other companies have seen something that the rest of industry, including Mobil, hasn't seen. Again, maybe incorrectly." He went off down the corridor. "And that," he said, looking back over his shoulder, "is what makes this a ball game."

Two friends call during the day to ask if I'm still alive. I say I am. The rig they heard about, the one that just sank, was the *Ocean Ranger,* a monster of a machine drilling for Mobil in the Hibernia area 175 nautical miles off St.

John's, Newfoundland. It was not *Midland*. I'm due out on her in a few hours, though, and I wish I weren't. It looks as if the whole crew, eighty-four men, went down with *Ocean Ranger*. That is not as bad as the floatel, the dormitory for drilling crews, that capsized in the North Sea, but it is the worst rig disaster in offshore drilling history. Rigs are like airplanes. They rarely go down, but when they do they can take a great many people with them. I call Mobil base to see if they have more news. No, says the woman answering the phone. "The oil patch is a small place," she says. "We knew a lot of those people. Do you still want to go out?"

"I guess."

"Well, you know, when your number comes up — you gotta look at it that way."

I drive down to the Airport Hilton. They have an illuminated sign out front: ASK ABOUT OUR HONEYMOON SEND-OFF PACKAGE.

The newspapers are full of the disaster. *Ocean Ranger* was the biggest rig in operation; its pontoons were about twice as long as *Midland*'s. And that's where the trouble seems to have developed. The speculation is that somehow the electronic ballasting system went out of control. The rig listed badly and lost balance. Former crewmen have been hunted up, and one is reported as saying the rig had started to list some days before the accident, but that trim was reestablished. Another went back to another marine disaster that occurred in roughly the same part of "Iceberg Alley." "We had a *Titanic* complex," he said. "We thought it could stand the roughest seas in the world." The rig had ridden out worse storms than the Valentine's Day gale. But now there were widows and orphans ashore, the biggest concentration of them in weathered old St. John's, and the town was passing the hat. When you have a *Titanic* complex, you don't buy much life insurance.

A few of the loungers in the Bristow waiting room are

reading papers. Most are sleeping. One Schlumberger man (the Schlumberger Corporation handles some of the well logging) says his buddy is pulling a lot of shore duty. "Now he can eat all the bugs he wants to." He fakes a retch. "Them lobsters."

Bob Garvin is his usual worrying self. They say the super he alternates with, Charlie Stevens, is just the opposite; he keeps things to himself. Either way, the well gets drilled. The CRT screen says the bit is at 10,892 feet. The temperature is climbing with the depth; it is up to ninety-two degrees Fahrenheit coming out of the well. Bob is in streaked-up, interbedded limestone and shale, and he's having trouble breaking in the new bit. Richard Walker is going back on the chopper for a week's work at Davisville. He has on yellow sunglasses and a natty gray going-to-the-beach jumpsuit with his name stitched in red over the breast pocket. He says it took a week to rig new anchor lines, repair the blowout preventer, and generally tidy up the place after the big storm. We're about eight days behind schedule now.

Bob sees my newspapers. "Don't read about the rig," he says. "What happened to the stock market?"

I start to say something and stop. Maybe oil men are like fishermen. Maybe the frontier is like the offshore banks. Where hunters work, you don't say much about those who fall prey. Bob finally does get around to it. "Could have iced up," he says. "Could have had a stuck valve."

I tell him the papers think the rig operators had neglected to observe some safety procedures.

"Look!" Bob does a little towering over me. "On this rig, we pay attention to safety. These are my people, and compared to them, one of these wells is not worth my little finger." I look up to see if Bob is laying one on me. He doesn't seem to be. He takes out after the press quoting people who used to work on *Ocean Ranger*. Some of them are bound to be sour, especially if they've been run off, he says.

The papers carry the rumor that *Ranger* hadn't properly tested her survival craft. Bob runs through *Midland*'s complement: a rescue boat, two survival capsules, and four inflatable rafts. He says they don't launch the bigger craft unless they have to. "Those cables are too slow for recovery. You'd bang your boats getting 'em back. But I've ridden them down, and they work fine."

Luck, says Bob. The two rigs near *Ranger* weren't hurt. "You never know. We lost a man off Alaska one time. Fell right off the rig into a rift in the ice. He should have been gone, but he came up in another rift yelling for the crane. We lowered the headache ball to him and he came up on it like a monkey fucking a football."

Richard Walker's relief comes up. His name is Murphy Comardelle, and he looks as if he just got his face back from the tannery. It is fine Irish hide set around a bayou mouth. Murphy accepts Bob's characterization as a twenty-four-carat coonass. Then he tells Bob the new bit isn't doing four foot an hour.

Bob snorts and yells, "Sonofabitch won't drill. What's the matter with it?" We go up to the drilling floor, Murphy and I, middling-size men, sprinting behind Bob's muttering hulk. The driller has his weight on the bit right. From there to the Exlog van for a look at the cuttings. Bob worries he'll have to change bits, and he's trying to figure out how to do it to cut downtime as much as he can. At this depth, tripping up to make the change is going to take hours. He decides to wait. "Sometimes, you just have to leave the bit down there until it gets to know where it is, and then it will commence to cut."

Minerals Management Service comes to call in a helicopter from Hyannis. I remember the men from a couple of months ago when they were out looking over drilling reports and poking around the rig to make sure *Midland* was conforming to their regulations and stipulations. They were concerned then that some of the hydraulic lines, put in for work in the Gulf of Mexico, might freeze up. Now they don't have much on their minds. They would have had more fun yesterday, Murphy Comardelle says. The roughnecks were "jumpin' lak beans" to keep from getting hit by chunks of ice falling from the derrick.

In the afternoon, *Midland* schedules a hydrogen sulfide drill. Sometimes, particularly in certain areas of the Gulf of Mexico, the bit cuts into a formation whose pores and crevices are filled with the gas. Occasionally, some of it escapes onto or into the rig. Chances are slim that it would happen here, but if it did, crewmen would have to move fast. A sensor would pick up the gas long before the rotten-egg smell drifted into noses. Sirens would go off and red lights would go on. "One of 'em let loose at two in the morning a couple of days ago," Bob says. "Scared the shit out of everybody until they found out it was a short circuit."

Hydrogen sulfide in fairly low concentrations causes nosebleed and vomiting. If things get worse, it goes after the nervous system and can kill you. A crewman shows me how to don the airpack you use in a real gas situation. It's a dry scuba. Slip it over your head and onto your back. Tighten the straps. Turn on the tank. Breathe in the mask. Tighten mask straps. Get out fast. "Make sure you got air," the crewman says. "I'm the one who'll have to come back in after you if you goof up."

Next morning, Bob is calculating his weights. *Midland* is carrying a lot of casing, and the deck load overall is creeping up on the limit of eighteen hundred tons. He is going to off-load some heavy stuff onto a workboat. "Trouble out here is," he says, "with the weather like it is, you have to keep everything on board you think you're going to need for three weeks." Deck load counts a lot if you get to rolling heavily. *Midland*, Bob says, took a couple of ten-degree rolls in the last bad storm, but that wasn't a list. *Ocean Ranger* reported a list of fifteen degrees before her radio went dead.

Midland's ballast room is run by an ex–Marine warrant officer. He has a console of lights and switches and current indicators, wind recorders, ice indicators, pitch and roll recorders. "We play put-and-take to keep it level," he says, showing me how he feeds water back and forth among the trim tanks in the pontoons. *Midland*, he says, was tested at a sixteen-degree list in the shipyard. "But the question is, where's the pipe? If you're out of the hole, that might be three hundred tons in the derrick. That might put you over."

We look out a portlight and see a white ship closing on us. If she were blue, she would be *Oceanus*, out of Woods Hole. She is white: *Endeavor*, the sister ship, run by the University of Rhode Island. I tell the ballast man she's doing some environmental testing for the government. He fixes me with a gyrene eye. "Half-assed environmentalists," he says.

In the afternoon, the sun comes out, an alien in this season. The rig idles in the sea in a rare easy time, making hole now at almost ten feet an hour. Roughnecks stand around on the drilling floor. Charlie Thompson, the driller, is the only one you can tell is concentrating. As the bit digs into the rock two miles below him, he lowers the hook to keep the weight where he wants it and then stops the hook's descent. The brakes scream. Charlie says there are a couple of roughnecks he's going to turn into derrick men. Let them play with the rig a little, learning how things turn to the right, then send them to training school. "It takes all your life," Charlie says, squinting up at the hook, keeping his dials at the edge of sight. "All your damn life to make a good driller."

Bob, tail in his swivel chair, is more relaxed than I've seen him. He is talking about his boys, how they went to school in Norway for a while, and how they are doing in football and baseball and shooting and the other skills he finds useful when you're coming of age in Texas. He expresses some satisfaction about the hole *Midland* is making. So far, they haven't used too many bits, and that's time and money saved. If *Midland* keeps drilling away, she'll catch up to Shell pretty soon. "Now that would be fine."

And then Charlie Thompson calls down and says he's losing mud pressure. "Oh, Lord!" Bobby Garvin cries, and his eyes wince and wrinkle shut. You can almost see his ulcer jump. Just that four-letter word. The shift in luck has caught him before he was ready with the others.

Bob thinks the trouble may be a crack in the drill string, maybe in one of the stabilizers that keep the string centered in the hole. They have failed before in this hole. Perhaps the string has broken at the crack. If it has, they'll have to send down a tool and probe for it. "You may see a fishing operation yet," Bob says as we repeat the familiar run to the drilling floor. By the time we get there, he has his face back in working order.

Thompson is tripping up. As we watch the stands come up and the roughnecks bulling them over to the racks, Bob talks blowout. The biggest percentage happen on trips like these, he says. If you do get a kick, you pump heavy mud under heavy pressure down the hole. The mud is designed to control whatever is forcing its way in from the rock. The intruder is then circulated out, carefully moved to special valves at the wellhead. If it is gas, you flare it or vent it. If it's oil, you leave it in the mud and dispose of it according to regulation: no dumping at sea.

The next day, Ed says he'll be calling Dallas to tell them about the stabilizer problem. The thing was cracked. "Did you put in a new bit when you were out?"

"Yes sir," says Bob.

"Just thought you needed some guidance." Ed chuckles, Bob snorts.

Ed reports back that Dallas wants to use a lot of collars

in the drill string. These are large-diameter joints that add weight where it's needed, particularly during this on-and-off bit performance. They can also create a lot of drag. Bob wants to use weight pipe of smaller dimension.

"Dallas wants it this way," says Ed.

"I've got so many bosses I don't know who I'm arguing with," Bob says. He keeps worrying at it over lunch. "I don't have anything left on deck to go fishing with if we twist off." But then Garvin the professional overtakes Garvin the galled. He checks his books and finds Dallas is right. The hole is too big to use the weight pipe he wanted. "I'd better call and apologize," Bob says.

When I drop in to say goodbye, Bob and the geologist are talking about drilling overseas. Both have another week to go on their tour. The geologist says that when he was on a platform off Africa, he used to kill time by running around the helipad, sometimes for hours. It got so he would fall asleep running, he says. "I was in danger of death by deplatformization."

Bob has the opposite worry, of time killing him. The company man and the rig super are always on call. They are captains. "You gear yourself to be ready," he says, looking at me with friendliness — I am, he told me yesterday, as squirrelly as anybody out here. "You take it easy when you're running well." The rig is tripping down, barely moving in the swell. "You unconsciously gear yourself for the number of days you're out here.

"Funny," Bob says, as the Bristow Sikorski shakes the air with its settling. "When you go a day over, you go crazy."

5

Georges Observed

"WAIT until slack water," the captain of a Woods Hole research vessel told me when I was a stranger there a decade ago. "Wait until then, and you've got half a chance." He was and is a cautious man and a good skipper. Others more daring or foolish will try the passage when the current is running full bore, when the buoys are drowning in the narrow channel that gives this town its name. Hang back or hell-for-leather, once you make it you're moored in just about the best deep-water snuggery in or around Cape Cod.

The haven and the teeming and relatively clean waters nearby are the main reason why the place is a world-class center of marine sciences. The United States has other concentrations of ocean academics, notably Scripps in California. There are first-rate facilities in the Soviet Union, Japan, and Europe; the French are especially intense about building their competence. But in Woods Hole, which is just a few square miles of promontories, inlets, and glacial mounds, you can count four year-round places of nonprofit business conducting marine research. Two are branch offices of federal agencies — the Geological Survey and the National Marine Fisheries Service — and the others are private — the Oceano-

graphic Institution, known locally as the Oceanographic
or just "Hooey," for WHOI (a source of amusement for
Russian scientists who used to visit during the last international thaw: what they heard approximated, in their
language, slang for the male member), and the Marine
Biological Laboratory. A seasonal fifth occasionally
glances seaward; about a mile up a lovely twist of road
from Eel Pond in the center of town is the summer sem-
inar center of the National Academy of Sciences, that
body of eminent individuals that reports frequently and
at length on scientific aspects of public policy.

In the summer, tourists come by the tens of thousands
to Woods Hole. Some are waiting for the ferries that will
take them to the islands: Martha's Vineyard, in sight
across Vineyard Sound; and Nantucket, almost three
hours' steaming to westward. A good many are down for
the day to see the ships and scientists. Maybe take a tour,
you know, and then find a nice beach. Parents in the
pinks and greens of vacation walk up and down Water
Street past the brick laboratories, the one drug store, the
small and high-charging food store, the handful of
crowded restaurants. Their children fret and dribble ice
cream on the sidewalk. Dogs, a disproportionate number
of them black, add to the press and the mess.

The trouble is that there are no tours. They would be
too disruptive of science. As for the ships, one, *Atlantis
II,* lies at the Oceanographic's dock, but she is inactive.
She is undergoing conversion to a tender for the institution's submersible *Alvin,* a tiny and tubby three-man device designed to work at depths down to 13,000 feet.
Other research subs can go far deeper; one has made it to
the depths of the depths in a trench seven miles down.
But the average depth of the sea is around 12,000 feet,
and that puts *Alvin* in more than sufficient demand.

The rest of the vessels are where they should be on this
average Oceanographic day: at sea. The disappointed visitor can check with the public information office at the

Oceanographic and find out where. *Knorr,* at 245 feet the biggest of the institution's vessels, is leaving Dakar, Senegal, for Recife, Brazil. She will be catching water again on this leg of her long cruise, taking samples containing minute amounts of tritium from the atmospheric atombomb tests and other chemical tracers that will help give a better reading of the oceanic circulation. *Oceanus* (the ship and the Oceanographic's magazine share the name) is down in the Caribbean acting as a platform for physical oceanographers making profiles of the many-layered sea. *Asterias,* a forty-five-footer, is working on a biological project off the Rhode Island coast.

The only thing left for a poor tourist to do is walk up to the Oceanographic's visitors center or walk down to the aquarium run by Fisheries, as the National Marine Fisheries Service is called hereabouts, where seals swim outside and striped bass inside. The bollards at the Fisheries dock are bare, too. *Albatross IV,* which often works out of Woods Hole, is heading north from Hatteras sampling plankton and fish populations.

No scientists are on the streets. Until the noon whistle lets go, spooking everyone, native and non, they are hidden away in their crowded labs and cubicles. Hundreds of them work at Woods Hole, along with hundreds more marine engineers and technicians. The biologists are the largest group, as they have been pretty much since the beginnings of modern marine science a little over a hundred years ago. Fisheries — the lab is known officially as the Northeast Fisheries Center of the National Marine Fisheries Service — is mostly fisheries biologists and technicians. At "Mumble," the Marine Biological Laboratory, scores of investigators — including, over the years, some thirty Nobel laureates — rent space for the summer (some are in residence year-round now) to work on research making use of the tissues and organs of marine organisms, from the axon of the squid to the eye of the skate. The Geological Survey is principally what its name implies. Only at the Oceanographic, the largest employer on the Cape next to the telephone company, is there wide diversity of disciplines: physical oceanography, chemical oceanography, biological oceanography, marine geology and geophysics, and ocean engineering.

At noon, some of these hundreds emerge. On summer days, the crowd on Water Street changes color, from the primaries to faded khakis and blues. Shorts, jeans, sandals appear, along with odd bits of gear and brown paper lunch bags. The shift is subtle and transitory. Within minutes, most of the monkish order is indoors again, munching sandwiches and listening to colleagues reporting on their work. These noontime talks are open to the public, but the topics often aren't: "Growth and Decay of Finite Amplitude Baroclinic Waves"; "Decapod Egg Attachment: Origin and Formation of the Outer Investment Coat." Among the munchers are scientists hard and soft — biologists and economists, chemists and anthropologists — who on their own and from their predecessors have gained more knowledge about more aspects of Georges Bank than is to be found anywhere else.

Fisheries was the first arrival. The lab was established in 1881 by a man with a shovel for a beard and a rare appetite for work. His name was Spencer Baird. Baird sold Congress on, and became the first (and unpaid) director of, the United States Commission of Fish and Fisheries. His job was to "ascertain whether a diminution in the number of the food fishes of the coast and lakes of the United States has taken place; and, if so, to what causes the same is due; also any and what protective, prohibitive or precautionary measures should be adopted." Baird dreamed of a biological summer university at Woods Hole, and in 1888, a year after his death, it materialized. Support came from Harvard, the Massachusetts Institute of Technology, a women's educational organization, and from wealthy men, not all of whom were comfortable with the direction biology was taking in those early Darwinian days. The story goes that one prospective donor, a Woods Hole resident, asked his solicitors if they were biologists. Yes. Did they believe in evolution? Well, yes, they did. "Well, I don't," snorted the prospect. "But I believe, gentlemen, that if you study diligently you will come to see your errors. Here is my check."

In a way, Mumble fathered Hooey. By the 1920s, when the National Academy of Sciences began looking for oceanographers in this country, it couldn't find enough to matter. One major reason for the scarcity was that no one was granting degrees in oceanography. Scripps was showing promise then as a training center for the West Coast, but what was needed was one for the East Coast. The director of the Marine Biological Laboratory chaired the study that made that recommendation, and the secretary of the venture was that remarkable moccasin maker, Henry Bryant Bigelow, who ended up running the brand-new Woods Hole Oceanographic Institution.

A fine brick building went up on Water Street in 1930, but the soul of the place came from Copenhagen. She was a big, steel-hulled ketch especially designed for oceanographic work, and for several years she was the only such vessel in the entire country capable of blue-water cruising. Her name was *Atlantis*. More often than not, she was called the A-boat. In her days of first encounter with the sea, mural painters, college dropouts, just about anybody doughty enough to try and strong enough to last were lured aboard *Atlantis* to flesh out the scientific parties. Some stayed on at the Oceanographic and a few have become chieftains in the clannish community of marine science. Susan Schlee, whose husband John educated the Senate panel on the possibilities of oil under Georges Bank, has collected some of the best A-boat stories and published them. About the lubberly lookout ("Land ho!" "Where away?" "Far, far away."); waters too deep for the hydrowire ("Damn, we missed the bottom!''); irreverent messages back to base ("Sea state, 6; sobriety, o").

The ship had no showers, and her sanitary facilities were the source of much inspired bitching. But she worked. She worked a great deal, with Bigelow's encouragement, on Georges Bank. She worked out on the Gulf Stream and down in the Caribbean, where just before the Second World War she helped the U.S. Navy improve its techniques of hunting enemy subs and protecting its own. In the end, when she was sold for a song to the Argentines after thirty-five years of service, after more than a million and a half miles of voyaging, someone found the original Danish bill of sale aboard. In the language of matrimony, it consigned the A-boat to the Oceanographic "to have and to hold . . . forever."

Bigelow went after Georges Bank itself, not just its fishes. The last of his three mammoth monographs describes what his measurements indicated were the temperature profiles and current patterns of that crazy water. He wrote of the sharp falloff on the seaward edge of Georges. He described the boat-killing shallows, Georges Shoal and Cultivator, and the way they seethe in the storms. And he was one of the first to study the gyre around Georges, the closed or partially closed clockwise drift so important to its biological productivity.

Lewis Thomas, surely one of the finest writers to come out of science in this century, has spent a lot of time in Woods Hole. He knows the Marine Biological Laboratory well. After a typical Friday-night summer lecture there, he wrote in *The Lives of a Cell,* there is a "jubilant descant, the great sound of crowded people explaining things to each other as fast as their minds will work. You cannot make out individual words in the mass, except that the recurrent phrase 'But look —' keeps bobbing above the surf of language."

Thomas also knows the myths in the public mind about the scientific method, about the precision of its reductionism, the rigor of its hypothesis. "It is not like that at all," he wrote in *Amicus Journal* (summer, 1981). "The scientific method is guesswork, the making up of stories. The difference between this and other imaginative works of the human mind is that science is then obliged to find out whether the guesses are correct, the stories true. Curiosity drives the enterprise, and the open acknowledgment of ignorance. The greatest single achievement of science in this most scientifically productive of centuries is the discovery that we are profoundly ignorant: we know very little about nature and we understand even less."

In many if not most branches of oceanography, verifiable stories of what goes on in the sea are relatively recent ones. It was fifteen years ago that a dean of ocean academics declared that marine science was on the verge of "making the silk purse of theory out of the sow's ear of observation." Even sows' ears are still somewhat in short supply, and small wonder. Landsmen may not realize that they inhabit by far the lesser part of this misnamed planet. The oceans cover more than two-thirds of it. About one and a third billion cubic kilometers of seawater lie there for our study. But the subject itself resists familiarity. Water is murky to the eye and to the high-frequency radio waves that have been such a boon to students of the atmosphere. Much of the work must be done mechanically or sonically, using instruments whose increasing sophistication in no way guarantees them immunity from corrosion, currents, or an occasional fish bite. Carl Wunsch, a marine scientist from MIT, wrote in *Oceanus* (volume 24, number 3), "It often comes as a surprise to the non-oceanographer to discover that the ocean is unobserved. With very few notable exceptions, there are almost no routine reports on the temperatures, velocities [and other properties] of the sea." The Gulf of Maine and its outer banks have seen far more scientific traffic than most stretches of ocean, but even here observation is irregular.

The difficulty often is synoptic, knowing what is going on at the same time over large expanses of sea. Only in that wide angle can the scientist understand the interrelationships of sun and sea surface; winds and waves; currents like the Gulf Stream eddies, those slow and immensely powerful storms in the sea that can disturb the patterns of life in their paths; pollution and production; reproduction and survival. Some help is coming from satellites, but so far their sensors aren't able to read much more than the skin of the water. Despite what Murray said to Bigelow about rowboats and bobbinets, the principal means of large-scale oceanographic observation remains what it has always been, what Carl Wunsch calls a "slow, expensive, labor-intensive, and often highly uncomfortable instrument": the ship.

The vessel specifically designed for oceanography is a recent phenomenon. Up until a half-century ago, most vessels used for research were yachts, converted or on loan from wealthy benefactors, or naval craft or fishing boats. The modern R.V. — for research vessel — must be able to carry a lot of scientists to sea in at least relative comfort. She must be sea-kindly, something less than a corkscrew in heavy weather, and adept at seakeeping — able to stay out when other ships run for harbor. She must be maneuverable in special ways, responsive to commands that will keep her riding over a bit of bottom no matter what wind and wave are doing. Her speeds must be governable down to tenths of knots, particularly at the low ranges needed for "flying" complex and expensive sensors at requisite depths. She must have versatile fittings, equipped with quick-fix bolts so that equipment may be secured or moved without undue fuss. She must carry an array of cranes to deploy and retrieve heavy objects, from current meters to cameras, and a crew expert in their operation.

No ideal ship yet exists. Many research centers, strapped in these two-edged times of increasing pressure for research and slackening federal funds (Washington provides about three-fourths of the Oceanographic's annual budget), have turned to smaller vessels for their coastal work. But the weather on the continental shelves is apt to be as bad as it is further out, or worse. Allyn Vine, an oceanographer at Hooey who has made a long and serious study of research vessels, says many of the new inshore boats "will scare you to death long before they drown you." For that reason, and for reasons of space to accommodate the interdisciplinary teams that make up so many of today's scientific parties, large ships like Scripps' *Melville* or her sister, *Knorr,* are often the preferred vessels, even in shallower water. They are costly. *Knorr* runs at over $11,000 a day, exclusive of scientific salaries. Even with their size, scientists often find them difficult to work on, particularly if the work requires long periods of concentration. Watches intervene. So do the noises, the thud of diesels and the whine of blowers and, at worst, the skull-busting rattle of the paint chipper. The Oceanographic's Henry Stommel, one of oceanography's most revered scientists, spends about a month at sea each year. "It's a little like going to a wild party," he says. "When it's over, you say, 'My God! I'm never going to do *that* again.' And then you see it's just enough fun so you do."

In size classification, R.V. *Oceanus* belongs somewhere between *Knorr* and the small ships that scare you to death. She is 177 feet long and, for that length, good in a blow. If you have the money to pay for diesel fuel, the price of which has been doing unspeakable things, she can cruise at fourteen knots and give you a cruising range of 7,500 miles. She can stay at sea for a month and keep her complement of twenty-four (evenly divided between scientists and crew) well fed and comfortable.

A partial list of her attributes would include two laboratories, one with running salt water; a single, controllable-pitch screw and a bow thruster for increased maneuverability; compasses, radar, speed log, depth finder, radio direction finder, loran C, satellite navigation system; side band and VHF radiotelephones; a hydrographic winch and a trawl/coring winch; a hydraulic A-frame; a fifteen-foot rubber boat and a twelve-foot skiff to help tend floating gear.

We are going out for nine days, out to Lydonia Canyon on the edge of Georges Bank, down to a station near Baltimore Canyon, and home. We are to recover three clusters of instruments mounted on huge tripods standing on the bottom; three current moorings, tethered meters measuring the flow at various depths; and seven surface buoys performing various functions. All these were put in place during earlier research cruises. We are to deploy fresh instruments. We are to conduct hydrographic measurements along several transects of the Bank and in Lydonia Canyon itself, using sensors with names like XBT and CDT. The first, the "expendable bathythermograph," looks like a midget bomb and falls to the bottom trailing copper wires thin as a spider's filament along which travel impulses that provide a temperature profile from surface to bottom. The CTD machine — for "conductivity, temperature, and depth" — is part of a washtub of a device lowered from the ship by the hydrographic winch to measure and catch water in sampling cylinders. Knowing temperature and density and other properties of the water, you can figure such things as circulation and the provenance of what you're measuring. We will also use side-scan radar and depth recorders to map the bottoms around the head of Lydonia. We will do all these things, says the preliminary plan for *Oceanus* cruise 88. Everyone knows we will not. Flanagan's law contains at least two premises. One, Murphy was an optimist. Two, if you accomplish half of what you set out to do at sea, your good luck will change. It is late in the year: October of 1980. Not all that late — the winter storms shouldn't be along just yet. But late enough.

The chief scientist is Brad Butman, a physical ocean-ographer who got his degree at the Oceanographic a few years ago and now is working for the USGS at Woods Hole. Butman is something of a rarity around here, hav-ing always worked in shallow water. He is conducting studies of sediment transport and circulation, partially funded by the Bureau of Land Management, which needs his data for the leasing and exploration on Georges Bank. His experiments have earned the highest accolade from scientists in Woods Hole years his senior: they are, they say, elegant.

Oceanus is tied up on the west side of the Oceano-graphic's dock, her fantail almost hidden under buoys, tripods, and "hard hats," the plastic covers for glass floats to be used in several arrays we will set out. Butman is poking around in the pile, waiting for one of his tri-pods — he did much of the designing — while a benthic biologist at the Oceanographic frantically straps instru-ments on the ungainly thing. The biologist wants the tri-pod deployed near a region of fine-grained sediment way south of Martha's Vineyard called the Mud Patch. He is studying, among other things, the role of benthic organ-isms (the tube-builders and other burrowers) in bind-ing the sediments, preserving them from resuspension by storm-generated currents.

The captain takes his ease in the ship's library — all the larger Oceanographic vessels have one — drinking his midmorning coffee. It is a pleasure to see him. Paul Howland was a mate on a cruise I went on not many years ago down in the Caribbean, and here he is in com-mand. An axe hangs on the bulkhead behind him, above a plaque: "When the automatic launching device froze at the crucial moment on the 19th of December,

1974, this broad axe in the hands of Dick Gabriel of Peterson Builders cut the safety rope and thereby averted the vessel's first shipwreck by allowing both ends to slide evenly down the ways." Howland invites me up to the bridge, well planned and well built, amidships, between the two stacks. He stands on the starboard wing to con us out. He calls the big island ferry in the slip astern to see if he's about to make his run. No, he isn't. All clear. Now we're creeping astern, and the big screw is kicking up the bottom. Howland sees Butman checking lashings and fiddling with a nephelometer — a device for measuring turbidity. "That man," he says, "is a fiend for work. The government gets its money out of that one."

A couple of weeks before the cruise, I had gone to see Brad in his office on the second floor of what was once a summer house on a huge estate overlooking Vineyard Sound. It is now part of the Oceanographic's holdings, rented to USGS. Butman must be in his early thirties but has the face of an uncommonly good-looking teenager. "A lot of my work on Georges Bank has been a huge data collection effort," he said, "cooperating with Fisheries and the Bureau of Land Management, and academic institutions like Hooey. The thing is, I don't feel comfortable with a lot of our analysis, and yet we're getting asked to answer all these questions right now: Where is the sediment going? Are the canyons important conduits for it? Will the drilling muds move and smother the benthos, the bottom life? It's just unfortunate that we're not a year ahead of where we are now." He laughed. "I suppose we'll always say that."

Paul Howland takes his time getting us out into the sound. There's a sill in the channel that doesn't have much more than nineteen feet of water over it at this point in the tidal cycle. *Oceanus* draws about eighteen, but we're loaded down. Out in open water, we go to eleven knots, a cruising speed dictated mostly by money.

Eleven knots uses a great deal less fuel than fourteen. Besides, Paul says, "she behaves better throttled down. She's good on the head seas, doesn't jump."

After supper, Brad goes over the cruise plans. We're behind schedule, so we're going to forget about making a dogleg down to a station south of the Vineyard and instead make straight for Lydonia Canyon. We'll get there around midnight and do some bathymetry, some bottom profiling, until daylight. Then we'll start recovery of what is on the bottom and deployment of what is on the deck. Brad turns to each scientist in charge of an aspect of the work — nutrient flux, suspended sediments, currents — and goes over the particulars for the next day. Then we break up.

First watch, four to eight. The graveyard hours. The ship is quiet, with only a medium whine from shaft and blowers and an underthrust of diesel. About half the watch is seasick; we're out of the sound and in the steep seas of the shelf. Mike Bothner, a chemical oceanographer Brad works closely with, goes about his meticulous work while carrying a bucket into which he can (as the Australians say) chunkygiggle, without missing a measurement. "Each time," he groans as he passes, "each time I say I'm never going back to sea, and each time I do." An electronic jumble lashed against the starboard bulkhead is leaking ozone, and the smell for some reason makes me think of frying grubs. We drive around over the head of Lydonia, transposing loran numbers to map numbers at two-minute intervals. Dawn shoulders up into the blue-black clouds. Someone is adjusting a recalcitrant instrument. "C'mon," he croons, twiddling knobs with both hands, "c'mon, machine." A land bird, a finch, is in the lab. So many of them wander out over the sea, so few find a ship. The knob twiddler captures the bird, and out he goes, in high protest, to trudge up and down the fantail, pecking hopefully at nothing.

Brad comes in to say a big low is on the way from the

Carolinas. We're going to need luck to do much of anything this trip. You can launch in rough seas, but recoveries are something else. It is no fun to snag a piece of heavy equipment like a tripod, which has to be recovered anchor and all, and snake it aboard when the ship is all motion. Better do as much of that as we can now. The call goes out to transponders on the bottom gear, and the signal releases retrieval lines carried to the surface by floats. We come alongside the floats, snag the lines, swing the tripods aboard. Mike Bothner climbs their slippery metal pipe to the top twelve feet above the deck to peer at his prized sediment traps. In them are layers of different-colored silts protected from the perturbation of benthic worms and bugs by a potent biocide. They contain fresh evidence that storms frequently stir up and resuspend the finer stuff below — important for science to know and for others who worry about transport of drilling refuse.

Brad steadies the tripod while Mike works above him. Georges Bank, Brad says, is probably eroding. Marine processes have been removing fine sediment from the bottom, leaving coarse material. The sediment on the deeper slope is finer than that on the Bank itself. He won't know until drilling gets along, he says, "but my best guess is that the very fine drill cuttings will fall to the bottom within some fairly close radius, say, a mile or so from the rigs. Then they'll be resuspended by storms and move westward along the shelf. The finest muds may end up in that mud patch south of the Vineyard, which is where we think many of the fine-grained sediments winnowed from the Bank have gone."

Mike swings down and heads for the lab.

"There are no natural sediments on the Bank as fine-grained as drilling muds," Brad says. "Just from that, you'd expect the muds would be winnowed away. The residence time depends on where they drill. The tides on top of the Bank are so strong that the muds might never settle there. Between sixty and a hundred meters, they may get to the bottom. But from the work we've done, I think that the winter storms will probably be strong enough to resuspend them and carry them out of the area in from one to five years. The muds may stay longer deeper than a hundred meters; we don't have enough measurements there yet to tell."

The ship turns to starboard, after another buoy. "We're just at the point of starting to understand the natural system and where natural sediments of a certain size are going," Brad says. "And we're particularly interested in the finer-grained sediments, since we know that oil and other contaminants adhere to them." The wind hits us with the turn, and we both pull our hard hats down over our ears. "The thing is," Brad says, "I don't have the understanding to say what the sediments introduced by man are going to do."

At the end of the day, the seas are eight feet high. The forecast says we'll get fifty-mile-an-hour gusts and twenty-five-foot seas when the storm arrives tomorrow. "Looks like an early winter," a crewman says. "January in October."

"We won't be able to work," Howland says to Butman. "If you want my advice, we ought to go back to Woods Hole and tie up for a nice weekend and then come back and finish this."

On a research vessel, the captain is responsible for the safety of the ship and those on it. What's coming doesn't seem too threatening in that respect. The chief scientist is responsible for getting as much of the research program done as Flanagan's law will permit. So we work through the night, launching tripods to clear the deck for the gear we'll be pulling from the water, streaming current meter arrays, each man glowing under the decklights in his flotation suit. Tag lines steady the heavy stuff as it goes up and over the rail. One-ton anchors splash over the stern and plunge to the bottom, and I imagine their instruments and floats swaying above them like outsize weeds. We watch the water sheave effect: the anchors on the current meter arrays sink so fast that their plunge sets up tremendous water resistance. The surrounding sea acts like a block over which the streamed arrays must pass as they follow the iron down. The big buoys at the end, designed to hold the mooring upright, race over the sea until they reach that point and then suddenly, and very dramatically, disappear.

The next day, the storm is off New York, and the pressure for recovery is stronger. The clouds are silver-gray and spit rain at us. We are now an ark for the land birds: nuthatches, finches, cowbirds, all tilted like old men strolling with their hands behind their back. A big heron flies up the starboard side and keeps going. A buoy pops to the surface next to the ship, brown-red with small life. "Sure is nice when things come up like they're supposed to," Brad says. They hook the buoy and lift the heavy vector-averaging current meter inboard and lay its long cylinder on its cradle. Two young women, biologists from the Oceanographic, pick animals off it — worms, amphipods waving brownly — and pop them in preservative. Polly picked a peck of pickled polychaetes.

Four women are in the scientific party, each with her specialty. One of them works with the gear, and she is expert at handling herself on deck. It's a long way from the days after World War Two, when a male oceanographer explicitly encouraged his colleagues to "think up reasons whenever possible to discourage women from participating in the work at sea," and advocated "an unwritten policy which does not prohibit but subtly discourages their presence" as the best way "to achieve our dubious ends." But there is a long way to go. Women receive only a little over 10 percent of the oceanographic doctoral degrees granted in this country. Since the general rule is that every woman must have a female roommate (there have been exceptions), there must be an even number on board. Even then, the old times linger. The story in Woods Hole is that on a recent cruise the chief scientist, a woman, was busily getting the research program organized when the captain appeared with a pair of his pants. They were ripped, and he wanted her to mend them.

The AMF Sea-Coder Model 200 B is acting spacy. The technician who serves it has been trying to interrogate a tripod that has been on the bottom since last May. Normally, if there is a "normally" in oceanography, the coder would chirp at the tripod and the transponder on the tripod would answer, all in a matter of milliseconds. The coder would also send a second signal through the water telling the tripod to release its float for recovery. "I talked to the tripod this morning," says the technician, "but it hasn't told me anything since then."

"Try real low power," Brad suggests. In shallow water, a lot of power can cause signals to bounce crazily from bottom to surface and back. The technician tries again. "Damn thing won't take the release command."

The instrumentation is complicated and a lot of it is one of a kind, but it is beautifully engineered and usually reliable, even after it has been in a bath of corrosive seawater for six months. The tripod might be bio-fouled, Brad says, deafened by a mat of growth. There is also a chance that it did release but didn't signal it had. If so, the buoy might be right under the surface, fouled, or held under by the tides. The moon is full, and the tidal currents could be playing the same game they would play with *Millie Bruce*'s anchor flags two years later as she tried to position *Rowan Midland*. Brad says we could wait for the tide to turn and see what happens. Or we could go after the tripod with a grappling hook. He's tried that sort of fishing with considerable success, and after a while he decides he'll give it a crack. The weather makes time as important as the tens of thousands of dollars invested in the tripod. Brad is pretty sure it won't come up unless he snares it. Working away on a wad of gum, trailing a spearmint wake, he goes to confer with the captain.

The waves are getting steeper and closer together. Down in the lab, someone has tacked up something from the Boston *Globe* on the bulletin board, a way to keep from losing your lunch. "Put a piece of tape over your belly button," the article says, "and keep facing into the wind." I'd rather not. It's over forty knots now, coming in from a little south of east. On Georges, a storm center usually lies full on your right hand if you're heading into the wind. This one is sneaking up between us and the mainland, and it is strong enough now so that the bow-thruster doesn't have the juice to bring us around by itself on the first try for the tripod. *Oceanus* hangs for a moment but then plunges away, trailing eight hundred meters of trawl wire and one four-point grappling hook. We make a curve across the wind and then drift down, hauling back. Brad is up on the bridge wing with Paul Howland, trying to plot the position of the tripod relative to the ship, trying to make himself heard in the wind rush. "Sometimes I think our equipment is in the Dark Ages," he yells at me. "I wonder if there'll ever be an easier way to do this." We circle and drift and play with the wire. Nothing.

Big rollers are coming in now, and the wind is beginning to raise the fur on their flanks. A seaman sprints out to close the vents for the clothes dryer. Howland throttles back and never takes his eyes off his man out in the storm. In the lab, a few scientists are playing Invaders on the Apple computer. The rest are out on the fantail, battening down. The sky looks innocent, light gray, as if it's going to clear. And then we spot the tripod float we've been after. Howland somehow manages to get alongside. The bos'n, big and old and solid, hooks it. In a minute, he's holding the frayed end in his hand.

"We cut it," Brad says. The tripod is still on the bottom. Brad wants to make another pass with the hook.

Howland is against it. "We'll be taking water by then," he argues, "and if we were hooked onto the tripod, we couldn't get out from under." Brad shows his bulldog side. Let's make a pass with the transducer, he says, to see if the tripod's still upright. If it is, the instruments are probably still working, and that's good to know. Howland agrees. In a while, Butman has information: his $75,000 baby is still on its feet. As we turn away, a migrating thrush bangs into the hull, breaking its neck, and falls into the sea. It rides surprisingly high in the water.

We can't work any more now. After supper, most people head for their bunks. Talk drops among those on watch. We are too busy listening to our own transponders, observing our own storm conditions, to do more than look at each other now and then, with pity and apology.

Around midnight, a grinding noise shakes me awake. *Oceanus* is laboring, shuddering. It sounds as if she has bent her shaft or lost her screw. But the noise stops and the ship regains control of herself. When I go on watch, just before a brilliant dawn, Howland tells me what happened. "It was gusting past seventy," he says. "I thought the stacks would go. Then it went quiet for a second, and we heard the life raft go over. We were lucky to get it." The noise I heard was *Oceanus* backing down for recovery.

A trawler appears for an instant, riding one of the high ridges of water that are running at us down the suntrack. Howland says that around dusk last night, we passed close to a Japanese longliner standing by her gear near Lydonia Canyon. She was over two hundred feet long. Paul says he has seen one like her take fifty swordfish on one set. The trawler is flung up again. Howland says some of Butman's tripods have been moved or toppled by trawl nets, even though he marks them with large surface buoys.

Sargasso weed floats by, Sargasso weed in late October. The bos'n says the water temperature is about seventy degrees Fahrenheit. We ran off Georges in the night to get into deeper and safer water, but we're not as far out as the Gulf Stream. We may be riding in a warm-core ring. Brad says that from one to ten of the eddies nudge the southern flank of the Bank in a typical year. When one arrives, he says, "you see a strong current to the east, caused by the clockwise circulation around it, for one or two months, and then the eddy will drift a little bit more to the west and the current will stop." Rings are hot topics for research. The Oceanographic and other labs are studying their dynamics — their formation out of Gulf Stream meanders and their gradual decay. And biologists at the Northeast Fisheries Center in Woods Hole are interested in what these travelling eddies do to fish. "Their currents tend to rip away a large amount of water from the shelf and bring it out into alien slope water," Brad says. "That exchange could have a real significant effect on nutrients and plankton and the biological dynamics of the Bank. There has been one suggestion that if you have a year where you have four or five major eddies nudging the Bank at the time of spawning, they may remove eggs and larvae from the Bank, thus making your year class worse than it might have been otherwise."

Brad has spent about two years researching and writing a paper on the mean circulation of Georges Bank. The current here, he says, has three components. The tidal one is by far the strongest, about a knot and a half up by the shoals. Superimposed on that is the mean flow around the Bank, the clockwise gyre Bigelow found and others at Woods Hole have studied for years. That moves usually at about a tenth of a knot. The third component, only recently the subject of systematic measurement, is storm driven. Those currents have been clocked at near one knot at the bottom along the southern flank of Georges and are probably much stronger near the surface. Then, says Brad, you have to add in internal waves, the commotion caused by waters of different densities coming into contact with each other so that waves, some of enormous height, travel along the layers and, occasionally, break. And you have to add in the eddies.

The tidal currents are also complex. A particle of water over the Bank moves in a narrow ellipse, north and west with the rising tide and south and east with the ebb, gaining speed as the sheet of moving water is forced faster over shoaling bottoms, slackening as the bottoms drop off.

Out of this swirl come hints, some of them quite strong, of what makes Georges Bank so productive. First, its waters are rich in nutrients like phosphorus and silicates, supplied in part by the cold oceanic water welling up from the depths onto the Bank and in part by planktonic organisms, whose excretions contain some of the nutrients they have ingested. Second, over much of the Bank, waters are sufficiently shallow so that sunlight — essential to the growth of phytoplankton, the tiny plants at the base of the food chain — penetrates all the way to the bottom. Third, tides and winds mix the shallows so that the surface waters, which otherwise would tend to lose their richness to grazing organisms, are constantly being replaced by richer waters from below. Fourth, the clockwise gyre, which appears at times to be a closed circuit, may act to keep nutrients and plankton — including fish eggs and larvae — captive in a productive environment far longer than is normally the case.

Paul Howland is taking stock of last night's damages. A sharp roll knocked the bos'n down in his cabin and drove him into something that punched a sizable hole through his ear. Another man scooped out his palm on a broken cotter pin when he was securing gear on the fantail. Paul himself has a finger pulped like a blood sausage by a chain tightener. He doesn't think it's broken, but the rest of us aren't too sure. Howland and Butman talk over the options. Brad wants to deploy one more tripod, the one we were waiting for back at the dock. Paul says no, it's roughing up again, and he wants to get the bos'n's ear tended to. That's his turf, and Brad goes along. We'll work tonight, head in, arrive at Woods Hole tomorrow night, then sail as soon as possible to finish the cruise by recovering instruments placed out on the mid-Atlantic bight, that great curve of coastal water centered on New York.

We work late. The moon sits on the water like a bright buoy. "R.V. *Moon*," says the mate and laughs. Scientists and crewmen haul iron like roustabouts. Brad has dark circles under his eyes now, and I wonder how much his drive has let him sleep. The happiest man aboard is a young scientist whose wife is going to the hospital tomorrow. She was going to have to face it alone, but not now. The bos'n wears a muff of gauze over his ear and a T-shirt reading "Official Mattress Tester for Worley Bedding Factory." He sits on the floor of the ship's library playing electronic chess with the mate.

Brad spends about sixty days a year at sea. The mate and others in the crew spend about two hundred. The mate looks in better shape right now. He is responsible for the ship during his watch. Brad is responsible for the science, every hour, at all hours. One responsibility is solid, lively, doesn't ship too much water. The other is questionable, patchy as plankton, a mess of numbers difficult to arrive at and more difficult to interpret. It will take Butman and his team a year or longer to work up what they're bringing home in shipboard measurements and magnetic tapes from the current meters and other instruments recovered from the bottom. A long time to synthesize bits of data into a fabric strong enough to make a true story.

Well into Nantucket Sound, we hear the Coast Guard calling for the ones who haven't come back from the storm. "Fishing vessel *Irene and Hilda*. Fishing vessel

Irene and Hilda." The calling goes on as we creep into Woods Hole. Then it stops. Nothing but the clicks of the gyrocompass in the blacked-out bridge as we slide home.

Oil is one of the most complex substances known. In fact, it is so complex that it cannot be said to be completely known. No single crude has been successfully broken down into every one of its tens of thousands of components.

That greatly worried a Swiss-born geochemist named Max Blumer, who studied petroleum hydrocarbons as pollutants in the sea, mostly out of Woods Hole, and on land. The biological effects of chemicals are related to the fine structure of those chemicals, Blumer wrote, and therefore all components of a natural organic mixture must be known before chemists and biologists can effectively predict its impact. Yet when science has tried to get at the fine structure of petroleum hydrocarbons, "each new analytical advance appears to have revealed a greater complexity . . . the explosion of knowledge has been paralleled by an equal explosion of ignorance." Blumer warned that while analyses of oil remain incomplete, "we must remain cautious in adopting tolerance levels. We need now to look for a transition to a more realistic study of nature that acknowledges the limitations of our present analytical powers and the gaps in our understanding."

Max Blumer died in 1977. The work he started is being built on by John Farrington, who came to the Oceanographic in 1971 to work for Blumer. John is almost forty. He was born and raised in New Bedford. He roamed the docks near his uncle's sheet-metal shop but didn't consider going to sea. Instead, he went to a small local college he could afford, but he says his grades were terrible and his interest in science zero. Then he met a professor of physical chemistry who gave him the challenge he needed. He went on to the University of Rhode Island's outstanding school of oceanography as a doctoral student. He met a young scientist there who intrigued him by saying, "I don't know a thing about the ocean. Maybe we can learn something about it together." They did.

John's face is full and slightly avian; there is something in it of Edward Lear's owls. When something excites him, he puts a peculiar kind of explosive energy into his words, real velocity. His colleagues think he has a lot of his mentor in him. Like Max, they say, he demands quality, and he speaks out. Unlike Max, who was almost shy in what he said, John fires off his opinions. And he is more comfortable with uncertainty. "I think it's a good idea to recognize the limits of our knowledge," he said one spring morning as we walked along the cool corridors of the Oceanographic's Redfield Building toward his laboratory. "But we shouldn't get hung up on it so we can't make decisions in the interim."

Somehow, John's percussive conviction is rarely perceived as arrogance. He has become someone sought after not only by scientists but by those who make profits from and policy for offshore oil. In one of his frequent appearances before congressional committees, Farrington delivered a thoughtful statement on his studies of the huge oil spill that in the summer of 1979 began pouring out of a blown well in Mexico's Bay of Campeche for three hundred days and drifted onto parts of the Texas coast. But along the way, he told the committee that it had received less than objective testimony from an oil industry scientist, a "very clever selective quotation" from the scientific literature attempting to prove that crude oil is less dangerous than refined oils. The trouble was, he said, that many studies of marine oil pollution had been of such poor scientific quality that their findings could be interpreted pretty much at will.

Then John touched off his main message. "Questions

about long-term fates and effects of pollutant compounds in the marine environment are being posed at a rate which is far greater than the increase in funds." He didn't want the government to expand what he regarded as its generally shoddy practice of putting together research projects "with ill-defined objectives" and then accepting proposals to undertake that research that often came from "incompetent or marginally competent environmental research companies which are designated as competent by 'authorized contract officers' in Washington who would not recognize a hydrowinch on a research vessel if they fell over it." The thing to do, he said, would be to give adequate funding to good academic, government, and industrial laboratories for proposals they initiate and carry out. That said, Farrington thanked the committee. And the committee thanked him.

Farrington's lab is crowded with long tables covered with an alchemic variety of glass tubing. He guided me past these bubbling transparencies to a machine slowly spilling a roll of paper on the floor. This, he said, was the first part of a two-stage process known as gas chromatography–mass spectrometry. What goes on here is a good deal like what goes on in a center for emergency drug treatment. At the center, a sample of the patient's blood or urine is concentrated and injected into measuring devices linked to a computer containing data on a suite of drugs and their metabolites. The drugs in the specimens are matched against what the computer knows. "It's possible under certain circumstances," John said, "to identify not only the drug but how long ago the person took it, and all that's flashed back to the technicians and then to the doctors so they can prescribe treatment."

With his equipment in Redfield, John's group can identify components of a crude oil or Number Two heating oil or viscous Bunker C fuel oil. They begin with a mystery of saturated hydrocarbons: polycyclic aromatics, some of them among oil's most toxic components; cy-cloalkanes; substances containing sulphur; substances containing nitrogen; and other assorted ingredients with names like ketone, phenol, asphaltene. They inject a carefully prepared specimen into a glass tube full of silica coated with a chemical film. The chemicals in the specimen behave in different ways during their passage through the tube, and their behavior is recorded as spikes and peaks along lines drawn by the gas chromatograph on the tongue of paper curling on the floor. If further elucidation is needed, the group heads for the mass spectrometer, which bombards the specimen with high-energy electrons. Again, the compounds react in different ways, with even more differentiation than in the chromatograph. These reactions are matched against mass spectrometer analyses of known substances. "Then," Farrington said, "the computer gives you its best guess of what the compound is you're looking at."

Having identified as much of their subject as they can, the researchers work in the laboratory and at sea to find out what happens to it in the marine environment. John himself has studied oil spills up and down the ocean, the Campeche hemorrhage and the wreck of the *Argo Merchant* near Georges Bank. On shore, he has been working with a group at the University of Rhode Island that uses large tanks, simulations of marine ecosystems, to study biological effects of pollution. His experience appears to have made a pragmatist out of him in everything except shoddy science. He is not against offshore oil drilling. What interests him is more specificity and less generalization in arguing the issue.

"The oil companies say it's better to spill crude than refined," he said as he turned out the lights in the mass spectrometer's residence. "That kind of statement is probably true, if you take all averages into account, but it really does depend upon the composition of the crude oil involved. And that composition can change, remember that, in the same day, in the same well. Some crudes look almost exactly like fuel oil in terms of their chemical composition. Hibernia, for example, that big find off Newfoundland. The range of boiling points in that crude runs right up to what you would find in Number Two fuel oil.''

We walked back to Farrington's office, and he sat down and leaned forward toward me, unlimbered. "If you average oil spills around the world, you come up with something that is less than earth-shattering. But that doesn't mean that a spill in a certain location won't be devastating to fishermen or the tourist industry. We know that oil under certain conditions and certain concentrations can be toxic." There's no argument against that, Farrington said. "That's clearly recognized by the industry as well as others. The real critical question comes in with the components you're talking about.''

Certain petroleum components of high molecular weight are not only toxic but last longer than other components. Aromatic hydrocarbons are one example. They contribute to the "smell" of kerosene bait that so attracts — and poisons — lobsters (fishermen used to soak a brick in the stuff until the practice was prohibited). Some components, like benzopyrene, last a comparatively shorter time but — either on their own or through metabolism or exposure to sunlight — yield carcinogens and mutagens. "So the question is," John said, "what are you worried about when you talk about toxicity? Effects on larvae of organisms that may live only for months? On whales, that may live for decades? Or are you worried about a poison that lasts long enough to affect us when we eat seafood? When you add it up, you're asking yourself the question, 'Okay, two or three years down the road, what are going to be the impacts of the oil you start with, plus its metabolites, its reaction products, on the success of a fishery like Georges Bank?' "

Farrington's face reddened with the effort to explain

what was on his mind to this nonscientist, this other. "Let's face it. If a large oil spill on Georges Bank was going to have the kind of impact of a volcanic eruption, there wouldn't be much controversy. People would say, 'Yeah, that's bad.' But we're not talking about fish being alive or belly up. It's something in between: whether there is going to be a long-term erosion of the fishery; whether we're going to wake up too late and find out there was something we should have done that we didn't do."

The topic of petroleum in the ocean has been one of the most controversial of any in marine science. The National Academy of Sciences has mounted two attacks on the issue. The first report of its findings appeared in 1975, after some delay while experts attempted to settle their differences. The second was scheduled to appear in 1982 but has been delayed, as of this writing, for the same reasons. John Farrington was a participant in the first round and a member of the steering committee of the second. Both reports deal with what the initiated call routes, rates, and fates. And with effects.

To the layman willing to dig in such literature, the surprising thing — surprising after all the alarums and excursions over offshore oil operations — is that rigs and platforms account for a tiny amount, perhaps only 1 or 2 percent, of the total amount of petroleum hydrocarbons delivered to the coastal areas of this country. Accidental spills of all types account, globally, for 10 percent. The operation of the world's seven thousand tankers contributes 14 percent. But over half of the petroleum hydrocarbons in U.S. coastal waters comes from what Farrington calls "the constant dribbling of oil into the environment as a result of our industrial and domestic use of oil on land": urban runoff, sewage outflows, wastes from refineries, dumping of crankcase oil down the storm sewer by your friendly neighborhood service station.

Scientists don't know much about the petroleum by-products delivered to the ocean through the atmosphere. They may be around 10 percent of the total delivery of petroleum hydrocarbons. What that may be is anybody's guess; the National Academy of Sciences has faced the same gross uncertainties in estimating the amount of oil in the sea as the United States Geological Survey has in estimating the amount of oil under it, and the best the academy could do was establish a range with an annual mean around six million metric tons. The largest source of the airborne component is incomplete combustion of automotive fuel, but there are others — smokestack emissions, for example. John Farrington told a congressional group that he and a colleague had found polynuclear hydrocarbon concentrations in ocean sediments deposited between 1960 and 1970 that were up to a hundred times greater than those in sediments laid down between 1800 and 1860. The implications are clear, Farrington warned. The hydrocarbons he had found weren't from oil spills but from atmospheric transport. "Even if we used up the available petroleum resources, the projected increased uses of coal and shale oil will result in a continued release of these compounds to the environment, which can have deleterious effects on man and on natural resource populations."

We are also pretty much in the dark about seeps. Almost two hundred submarine seeps of oil and gas have been identified around the world. They occur with some frequency off California and in the Gulf of Mexico. Many remain undiscovered. A research vessel recently found a huge lens of oil two hundred meters down, way off the northeast coast of Venezuela, but there was no subsequent contact. Popular science has a high old time with the phenomenon; one theory has it that the disappearance of planes and ships in an area known to sci-fi readers as the Bermuda Triangle is due to great belches of gas from the seafloor that discombobulate vessels and smother engines in flight.

Seeps from the estimated 2.5 trillion barrels' worth of petroleum resources under the ocean floor may account for 10 percent of the hydrocarbons injected each year into the marine environment. They pollute, yes, if you stretch the term to take in the harm nature does to itself. But those few scientists actively studying them find there is no devastating damage, that often there are plentiful and varied benthic communities living around them. John Farrington says one reason this is so may be that seeps usually are long-term phenomena to which organisms can adapt. Another is that "much of the seep oil is weathered by subterranean contact with seawater. There is some fresh seep oil, but a lot of it has lost its original — and its more toxic — components."

What happens to oil when it gets into the ocean is a matter of variables dancing with variables. The fates of petroleum hydrocarbons depend on their composition, the nature of the patch of ocean into which they've been introduced (the temperature, the salinities, the wave and current patterns), and the type of ecosystems at risk (intertidal, subtidal, estuarine, or open ocean). Not all oil is spilled on the surface; the IXTOC blowout near Campeche spewed it directly into the water column from the wrecked wellhead, and it surfaced miles downcurrent. But fresh oil tends to float. When it does, and when seas are calm, it tends to spread out over the surface in a slick, the thickest portions at the leading edge and a wake of sheen trailing out behind.

From a third to two-thirds of the floating mass is usually lost to evaporation within a few days. Lost is not quite right: components of the oil can return to the sea later and often in more toxic form, thanks to the photo-oxidation effects of sunlight. Under certain conditions, the weathering oil on the surface turns to thick, sludgy pancakes. Under others, water mixes with the oil to form mousse, a noisome brown topping so substantial that boats sent to investigate islands of it floating away from

IXTOC just about went aground in the stuff. Mousse was thought to be relatively harmless, as floating oil goes, but researchers have found that it can insulate some aromatics and other compounds that would otherwise weather into harmlessness, carrying them for hundreds of miles in near pristine and poisonous condition.

Eventually, what survives from a spill at the surface is a collection of lumps called tar balls. They are spotted most frequently in midocean these days, bobbing along in or downstream from the tanker lanes. Earlier on, some of the spill went to ground: some settled on the surfaces of fine grains of matter in the water and sank, and was ingested by organisms and excreted as fecal pellets. Whatever the process, enough oil can end up in the sediments of the seafloor to be of concern. In time, even entrapped oil will be degraded, by microbial action or other processes. But John Farrington and his coworkers at the Marine Environmental Research Laboratory in Rhode Island have found that oil in a simulated marine environment can harm life on the bottom.

Time is the slippery customer here. Oil producers have been known to say that petroleum hydrocarbons produce no irreversible effects in the sea. Farrington applies his ultimate epithet to that argument. "Ridiculous! You've got to put a scale on the term and a scale on the effects. If it takes ten years to work through a serious impact on Georges Bank, that's reversible for the ecosystem, but it may ruin the fishery."

So what will oil or drilling muds or the other discharges and dribbles from offshore operations do to what lies on or swims over Georges Bank? That central question, asked in a hundred ways, puts the spurs to scientists like Brad Butman and John Farrington. It can still be answered only peripherally. The answer depends, the scientists say, on the type of oil or mud discharged, how much is discharged and over what period, the season of the year and the weather, the kinds of creatures in the way. Scanning the literature for evidence of impact produces much more of a soup than a synthesis.

Starting at the base of the food chain, there doesn't seem to be much evidence that oil pollution has seriously affected the mites of the sea. Severe oiling can swamp even the lipophilic bacteria that degrade petroleum, especially if the ambient seawater is low in nutrients. Extreme conditions can produce changes in the type of microbes present in a given area and thus changes in the predators that feed on microscopic life. Phytoplankton have registered reductions in their rate of photosynthesis (the elegant process of using sunlight to produce food for the photosynthesizers) that can be traced to oil. Sometimes pollution produces a bloom or explosion of life among the specks of sea plants, but this may be due to a falloff in grazing by herbivorous zooplankton killed or narcotized by the pollutant.

Loss of plankton to oil in the sea is extraordinarily hard to document, given carcasses usually far less substantial than that of a midge. And until quite recently, loss — lethality — was the only yardstick in use. Now, however, scientists working in laboratories can produce effects that stop short of death but that can be enormously important. In those experiments, copepods, the ubiquitous micro-oarsmen of the ocean, have slowed their rate of feeding in oiled water. Fish eggs and larvae are among the floating or weakly swimming planktonic hosts, and they have shown changes when touched by oil. The data are sparse, too sparse, some say, but after the *Argo Merchant* went aground, about a fifth of the cod eggs and about half of the pollock eggs collected at sampling stations were dead or suffering chromosomal or other damage. Polycyclic aromatics in petroleum are by no means the only substances in the sea thought to cause cellular problems, mutations, or cancer. But they are among them.

Adult fish seem to be more sensitive to intense and

sudden pollution than invertebrates, particularly if they are already under stress due to abnormal temperatures or salinities or scarcity of food. Pelagic fish, the endurance swimmers, are likely to be more tolerant of oil fractions in the water than their benthic brothers. Under the right conditions, oil can taint fish — as little as ten parts per million can do the job — or affect reproduction; the drop-off in some flatfish populations after oil from the *Amoco Cadiz* drenched the French coast is thought by some scientists to be attributable to that. Oil can affect feeding behavior, it can bring fish belly up, or it can do none of these things. There is evidence that adult fish avoid spills and evidence that they do not.

Most invertebrates, of course, either can't move or don't move far enough to avoid the consequences of petroleum pollution. Oil in sediments in laboratory tests has caused crabs to have trouble finding food and clams to expose themselves to predators by burrowing less energetically than they normally do. The real crazies, though, are the lobsters. At Woods Hole, a young Dutch biologist (and flautist) named Jelle Atema has been subjecting *Homarus americanus* to various levels of petroleum and drilling-mud pollution. In both instances, he found that the animals' behavior changed, sometimes in the direction of acrobatics, sometimes in the direction of narcosis and death. Baby lobsters had a devil of a time digging down through a layer of drilling mud that had settled over the sediment at the bottom of the tank to a depth of only a quarter of an inch. "Although scientific work has not yet been done in the open sea," Atema declared, "such experiments can give some idea of how successfully exposed animals can cope."

So they can, but only to a point. The difficulties of reconciling lab work with sea work are enormous. Brad Butman's experiments, for example, show that muds probably wouldn't stay long on the more energetic parts of Georges Bank, though they might settle and stay on deeper bottoms where currents are calmer — and where lobsters also live. All Atema and other laboratory scientists can do is stand at the shore and point to potential dangers further out, like the thin man in the old *Philadelphia Enquirer* ads. Their findings may apply in the sea. They may not — and other as yet undiscovered dangers may. Even something as basic as how oil is mixed with water in a laboratory tank may cause the results of an experiment to go wide of the natural mark.

The birds and beasts of the ocean also come into contact with oil on occasion. Birds can suffer death from exposure and other stresses as the oil fouls their plumage and ruins their insulation. They can ingest petroleum hydrocarbons while trying to preen, but it isn't yet clear what effect that produces. Less is known about mammals. Rights, humpbacks, sperm, and other whales still scull along Georges despite the long years of hunting them, and seals live nearby. Analyses of oil impact on animals such as these usually come after the fact: inferences from autopsies. Some oiled seals and other marine mammals have been sighted, but no one knows their fate, or if their comrades swam into the slick with them or swam off. There are no reliable reports of whales getting themselves oiled. Polar bear kidneys and brains have shown heightened hydrocarbon content, which may be due to grooming after oiling, but the long-term results have yet to be determined.

Man himself does not yet appear to be much at risk from the substances he sends to sea. Tainting of fish by oil and other pollutants can raise hell with the income of local fishermen for a while. The threat of carcinogens and mutagens in oil ingested by organisms that in turn are ingested by us does not now cause great concern in scientific circles. All in all, the evidence is that natural carcinogens in our diet far outweigh anything added by man.

Copepods, cod, shearwaters, whales — the tendency

until recently has been to concentrate on what pollutants do to each in its own habitat. But the truth, if it is extractable, seems to lie more in complexity, more in ecosystem than in individual organism. Ecosystems are mystifying fabrics, even on land, where science can stand and see. Coastal ecosystems are far more difficult to study. Open-ocean ecosystems lie at the far edge of competence.

⁓ One of the most thorough, and certainly one of the longest, studies of the effects of an oil spill on a marine ecosystem was carried out by Howard Sanders, a senior biologist at the Oceanographic, near West Falmouth harbor a few miles up the Cape side of Buzzard's Bay from Woods Hole. Sanders is a mild man for whom precision in numbers, in data, is perhaps a bit more important than breathing. But once he has his numbers, he is apt to be a bulldog in arguing his interpretation of them. He arrived on the scene in the late summer of 1969, just as oil began drifting onto the marshes and beaches from a barge on the rocks offshore. He and his colleagues kept measuring for years afterward. They reported that many fish were killed immediately. Bottom dwellers died and were replaced by opportunistic species. Some marsh grasses were killed and others showed evidence of lingering damage. Sanders' insistence on meticulousness, on checking and rechecking, drove him and others almost to the breaking point. His findings were sensationalized by some environmentalists and ridiculed by some scientists close to the oil industry. But as other evidence from other spills began to build, it became clear that oil in inshore waters can have serious and long-lasting effects on plants and animals of the coast.

Some studies further offshore have also indicated changes, perhaps the most extensive being the one carried out in the North Sea after the Ekofisk well blew out in April of 1977. Work in the Gulf of Mexico, funded by the oil industry and touted as the "most comprehensive study concerning the effects of chronic exposure of oil to marine life ever attempted," showed, according to that same touter, an official of the American Petroleum Institute, that such low-level chronic exposure "has, at most, negligible effect on marine life." Some scientists, including Howard Sanders, went through the data and found that several of the study's investigators had been unhappy with its procedures, that control sites were not well situated, and that the whole study area, far from shining with health, was slightly but surely polluted. More money was pumped in and a second report issued that John Farrington says was much more balanced. Meantime, he says, glossy versions of the first and faulty summary report were sent to congressional offices. "The other versions," John says, "with the caveats and more detailed information, didn't make their way through."

Howard Sanders' report on the West Falmouth oil spill is now widely accepted. The danger of oil pollution at low concentrations over long time periods is gaining acceptance. "It takes time," Farrington says. Part of the delay in finding common ground can be laid to the natural resistance of opponents in any argument to shifting ground. Part is due to the difficulties of differentiating between what is known and what is suspected. And part is due to the nature of the research on oil in the sea: much of it is crisis research.

In December of 1976, I sat in what had been the director's office at the Oceanographic, listening to a bunch of keyed-up scientists and government people trying to work out a plan to study the spill welling out of the hull of the Liberian tanker *Argo Merchant,* aground in heavy seas off Nantucket. Dozens of federal, state, and local agencies were involved or should have been, and several research institutions. Where to find a ship? What gear to put on her and what scientists? Where to get the money? For hours, the group sat tending to logistics — and to egos. An emergency strike team from the National Oceanic and Atmospheric Administration was out at the wreck with Coast Guard specialists trying to figure out if the hull would stay together. It didn't, and the oil came out into the sea in gouts.

By purest luck, *Oceanus* was working in the vicinity, four hundred miles away from Woods Hole, when word came to hump for home. Plans for a "crisis reaction cruise" were ready when she docked. Men raced to unload her and reload with the instruments necessary to measure rates, fates, and impact of *Argo Merchant*'s nearly eight million gallons of Number Six fuel oil. *Oceanus* sailed the same day she came in. On board were scientists with familiar names: Howard Sanders and John Farrington. John Milliman of the Oceanographic's geology and geophysics department served as chief scientist while conducting his own research.

They did not know where the oil would go. "If it continued to float," Milliman wrote later, "presumably it would follow the direction of the wind, generally to the east in winter months. On the other hand, if it sank, it probably would move with the subsurface currents in a westerly direction." One study Milliman knew of showed that oil off Alaska had sunk quickly, sorbed onto particles in the water. The waters off New England tend to be less turbid, but Milliman felt the same thing could happen out on Nantucket Shoals.

The wind changed, and the oil began to drift eastward, out to sea. *Oceanus* steamed ahead of it, right into a storm. Milliman and his people were able to occupy two stations and do half the work called for — collecting preslick background information — before being forced back into the relative shelter of Nantucket Sound. The forecast sounded so bad that the ship returned to Woods Hole for Christmas. The cruise had been unsuccessful: not enough samples in the right places.

Shortly before the second cruise, the wind swung around to the southeast for some fifteen hours, and the slick headed back toward Nantucket and the marshes of southern Martha's Vineyard. *Oceanus* headed out to sample areas west and south of the wreck, mud bottoms potentially more vulnerable than those covered with sandy sediments where oil would either be washed away or oxidized. Scientists checked the rate at which oil was settling from suspension. Not much was. Then the slick turned again, and again put out to sea in a formation of sheens and pancakes. Those who thought all the commotion over *Argo Merchant* was so much hysteria had a field day.

Things could have been otherwise. Those southeast winds, freaks though they might have been, could have lasted long enough to do damage. The oil could have sunk and resurfaced close to New England bays and beaches. The wreck could have occurred during spring or summer, spawning time for many of the fishes of Georges Bank. But the gods were kind, and some painful lessons were learned. One was that oceanographers did not have enough experience in crisis planning to maximize research opportunities in the confusion of sea and ships and exhausted men. And there wasn't enough advance information available to begin to predict how the oil would act. "With increased federal and state-supported research," Milliman wrote, "such data might be available before the next disaster." Perhaps. The image I carry from my watch on the rim of the crisis is a photograph of a Navy diver being debriefed after swimming under and into the *Argo Merchant* mess. He is sitting on the deck of a cutter, oil running down his suit, down his faceplate. I have never been able to figure out whether he was grinning because he had made it through the dive or grimacing at what he had just seen, smelled, and felt.

It is one thing to work in a lab or on a fantail, subject to the wind, the weather, and peer review. It is another to go before congressional committees or mix it up with interest groups that don't like your findings. John Teal knows both worlds. Teal is chairman of the Oceanographic's biology department and husband of anthropologist Susan Peterson — she who studies fishermen. He appears shy behind his bush of beard. Yet his look is direct, at times hawklike, and the wooden clogs he wears, even in winter, mark him in any shuffling crowd. He is best known among scientists for his work on the natural budgets of marine wetlands, their productivity and decay, and among laymen for his book *Life and Death of the Salt Marsh,* written with his first wife, Mildred. But he often works in blue water; he was aboard *Oceanus* when that ship was tracking the oil from *Argo Merchant.* "I can call myself an ecologist," he says. "I can call myself an oceanographer." He laughs a little at himself, a marine man born and raised in Nebraska. "In certain audiences I can call myself a kind of chemist." He could also call himself a scientist in the ring. His kind of unruffled pugnacity lends itself to the long and often nasty skirmishing that confronts the researcher who walks into the arena of public policy and stays in.

We were talking about that kind of alley fighting in Teal's tiny office in the Redfield Building, a box looking out on the patchy flow of early-summer tourists along Water Street in Woods Hole. "The difficulty," Teal said, "is in making people understand the limitations of what we can contribute. We rarely have a very good ability to predict the consequences of one particular human action, and the ability shrinks the more complex the system is we're dealing with. But competent scientists can still make a better guess than people who don't know anything about it."

The phone rang. Teal is one of the most tied-down scientists I know. Everybody is after him for this or that, and he seems to pick his clients out of a bingo cage. He has worked with federal scientific review panels. He is on the board of the Conservation Law Foundation. He once advised a local developer who was worried the town of Falmouth might frown on waste discharges from his condominium project.

Teal hung up. "I was saying?"

I told him.

"Oh, yeah. Well, I am personally convinced that it is always better to act on the basis of a little knowledge than none at all. But scientists have opinions, and very often they don't make it clear where their expertise stops and their extrapolations begin. We all do it." He swung around and began talking to the window again. "The other problem is how the recipient of your information interprets what you're saying. If you're saying something won't happen and it does, how does he react? Does he realize you were trying to help him by extrapolating beyond your data, or does he think you've done a poor job as a scientist? It's a very tricky business. It's easy to get the feeling that it is so difficult, that being tripped up is so inevitable, that you would prefer not to be involved in policy issues."

Teal said that the pressure has forced a lot of scientists out of the debate and kept others from entering. "Take Max Blumer," he said, referring to the pioneer in marine pollution research whom John Farrington had come to the Oceanographic to work for. "Max was heavily involved in the controversy over the effects of oil, the dangers of hostile effects; and he got jumped on so heavily for some of the extrapolations he made that he decided to pull out altogether. Others have become defensive under fire. When that happens, you can retreat, you can lose your objectivity and then your credibility. When that happens, a scientist destroys himself as an effective —" He checked himself. "No, that's not necessarily true. Some people who have lost their scientific credibility have become very effective in their policy utterances, by virtue of the fashion of their address. That's even worse."

A young woman knocked on the door and lifted a thick stack of computer printouts to the tiny window for Teal to see. He motioned her in. "You can sympathize with people who want to pull out," he said. "I mean, somebody recently wrote about me, 'I don't see how any responsible scientist could have said . . .' " He barked a laugh and took the printouts from the woman. "But pulling out or never going in doesn't help contribute to solutions of public policy based on good scientific information. If you leave the politician to make the decision in the lack of science, he will — of course, he has to! And then, it seems to me, the scientist has nobody but himself to blame. You haven't got really much of an excuse for complaining about what is happening in society if you don't contribute."

The game is rarely played off the front wall alone. The angles of argument are such that a scientific statement made in a published paper or before a subcommittee may bounce back at the scientist via industry or an environmental interest group. Howard Sanders thinks he took a good deal of abuse over his report on the West Falmouth spill. Some industry sources felt that the spill was anomalous, that fish mortalities were caused by storm conditions. In a defense of his work, Sanders wrote that his most persistent critic had charged that the West Falmouth study "was probably the least competent of all analyses of oil spills," and had added that "oil causes little serious threat to marine life." According to Sanders, a technical magazine that wanted to publish a competent review of the matter had a difficult time trying to find a scientist willing to write one. The reason, Sanders quotes the editor as saying, was that a reviewer who criticized the critique of Sanders' study "might have difficulty securing research money from the two largest sources — oil companies and the federal government."

Sanders complained that "it is exactly this sort of thing that makes us fear for the integrity of science in the face of considerable economic and political power." At fault, he said, were the gray literature that "circumvents the route of critical peer review that science traditionally demands to protect its integrity"; the uncritical acceptance of such literature as a basis for public policy; "and a government that appears . . . incapable of correcting those abuses."

Sanders is not like Teal or Farrington. Some of his friends say that he charges like a knight on the road, that he may be right but he is often repetitious. Sanders' reply is that if a scientist working in a field like his decides to enter the public forum with his findings, he will encounter such powerful and well-organized opposition, with such ready access to the public ear, that he "has to present the problem in such a way that it *must* be faced." Because there are no checks and balances in the argument, he told me once, "my position *seems* extreme."

The gladiatorial approach deeply offends Sanders the scientist. His battles, he admits, have been at considerable cost to him, his life, his work. Without tenure at the Oceanographic, he says, he could not have taken the chances he has taken. "Clearly, a young person can't do this kind of thing, if he wants a career. There is so much effort involved that if you aren't careful you don't have time to be a scientist."

John Teal is more pragmatic. He joined a group of environmentalists and industry scientists that worked with the federal government in developing the interagency plan for monitoring the effects of drilling on Georges Bank. "I have a hobby horse," he told me before the drilling started. This was a chance, he said, "to organize a study in a place where there hasn't been any drilling. Drilling for oil is a marvelous opportunity for doing a large-scale scientific experiment on Georges Bank. It is going to do certain things. It is going to modify the environment in certain places. It's going to do it on a big enough scale so you can hope to measure the consequences and whether those consequences matter to the fishery. The oil companies say they support this experiment because they want to show once and for all that oil drilling does not hurt the environment when properly done. Other people start out with different biases, of course. My bias is that I think it would be interesting to get some good information."

As drilling got under way, Teal continued to push for the monitoring program. When it came, he began to look forward to a find on Georges. "In the scientific sense," he said, "I hope we find oil out there, because that will make the study worthwhile." That would provide measurable perturbation of the ecosystem.

John Farrington has Howard Sanders' passion for precision. He shares Sanders' contempt for the shoddy science generated by the federal government's demand for quick answers to environmental problems often best studied by methods that are labor-intensive and therefore slow. But Farrington the organic chemist is apt to be less accusatory than Sanders the benthic ecologist of research on the other side of the fence. When I asked John to compare oil company and academic scientists, he said there is no schism between the two. "There is a range on both sides. You know yourself that there are some very conservationist-minded scientists here at the Oceanographic and some that believe you can't get upset every time a couple of polychaetes happen to roll over and die. Industry scientists aren't constrained from submitting papers to the best journals, with peer review, and some have published very good papers. Their companies, and the American Petroleum Institute, have supported a large number of good research projects and publications." Then he paused and leaned forward. "But in the end," he said, "there is the question that lies at the base of environmental risk assessment right now: who has to

prove the existence of risk? Is it the person who's intro-
ducing the substance into the environment, or is it the
society that may benefit from the substance while being
harmed by its discharge onto the land or into the sea?
The political process hasn't resolved that yet."

＼Farrington believes that environmental forecasting for
places like Georges Bank needs to be taken seriously. The
process has its limits, he says, particularly "if people ask
questions requiring sampling so extensive that the structure being sampled is changed, or so long-term that by
the time you finished, things would be a lot different
from when you started." But he is optimistic. "I think the
probability is quite good that we will be able to give a reasonable answer as to whether a given environmental effect will be devastating over a ten-year period," though
finer gradations may exceed the system's capacities.

What worries John is that neither the government nor
the public is ready to give ecological forecasting its due.
"In trying to arrive at a reasonable decision," he told a
congressional hearing, "government officials are presented on the one hand with solid economic arguments
that such and such a benefit with so many jobs, and such
a multiplier effect on the economy of the region will result; and on the other hand with a set of 'possibilities' and
unknown risks presented by scientists conducting research on . . . long-term effects of oil in the marine environment. All too often, the 'solid' economic arguments
win out over the . . . scientific assessments of short-term
and long-term risks."

John Farrington, his face reddening, caution in his
back pocket, asked why this should be so when economists every day do such a shaky job of figuring out what
the economy is going to do next and why. "This," he said,
firing each word, "is a ridiculous situation."

She looks like a misplaced bridge. When you get closer, you see that the steel arches are attached to pontoons. They are old Navy pontoons, all but submerged. Her deck is mostly grating. There are rectangular vans outboard, forward and aft, and a box of a pilot house built over the stern arch.

Lulu? Lulu is the name of this jumble? Yes, but laugh not, for motherhood is involved here. Lulu was the given name of the mother of Allyn Vine. Allyn Vine is a scientist, now retired, at the Oceanographic. He is also one of the most persistent and respected students of the need for and design of research vessels of all types, including the submersible nurtured by this seagoing span. As Lulu mothered Dr. Vine, *Lulu* mothers *Alvin,* or will until her replacement, *Atlantis II,* is ready to assume the responsibility.

Alvin sits in its cradle forward of the bridge. The cradle is a crude elevator lifted and lowered by four Promethean chains on sprockets. The sub rocks and bows on this windy morning. We were late leaving Woods Hole yesterday, thanks to a last-minute paper chase through the front office to sign on two new crewmen, so we are still two hours from the dive site. *Lulu* can make six knots in a flat-ass calm, and today the waves over the outboard edge of Georges Bank are running five feet and building. The larger ones throw gouts of seawater up through the gratings to be blown to diamond spume.

Alvin faces forward in its cradle, and the scientific party faces aft, looking it over. This cruise is being financed by Fisheries, which will pay $14,000 a day for ten days. That will cover one day steaming from Woods Hole to Oceanographer Canyon and one day back and, weather willing, eight dive days on a small piece of incised continental slope not far from Mobil's first well. Even in placid August, the weather on Georges is never willing for long.

The cruise is part of what the National Marine Fisheries Service calls its Ocean Pulse program, a long-term study of changes in the sea attributable to human doings. Pulse stations were initiated in heavily polluted areas like Raritan Bay in New Jersey or the New York bight, where the Big Apple dumps its wastes. But in the midseventies, scientists at the Northeast Fisheries Center recommended and got a pristine station — if anything off the northeast coast can be called pristine any more: Oceanographer Canyon. "To make a long story short," says Joe Uzmann, "the place hasn't been messed up." Before anything can happen to change that, Fisheries wants to study the organisms in the canyon and on the nearby slope and see how they interact with their habitat; in particular, how they contribute to the sedimentary processes of erosion, transport, and deposition.

Biology and geology are linked on this cruise. Joe Uzmann, chief scientist, is a senior fisheries biologist with the Northeast Fisheries Center. He has a northern face, fair skin, and is not an early-morning person. It takes him a while to get started, he says, squinting in the spray and the sun, holding a cigarette and, an hour after breakfast, his first Coke of the day. It takes him longer today, because *Alvin*'s chief pilot has just told him that dive one is scrubbed. The wind is with the seas, shouldering them up, and that's the way it is supposed to be all day. Launch would be possible, but recovery would be just too dicey.

Joe takes a mournful pull on his cigarette and tosses it down through the grating. "Well," says his second in command, a gentle geologist from the Geological Survey named Page Valentine, "there's always bathymetry." Page cleans his glasses, wipes salt rime off his Grouchoid moustache, and heads for the science van, up forward over the starboard pontoon, bound for some bottom map-

ping. Joe and the rest of us follow: three men in their
thirties and forties from the universities of Connecticut
and New Hampshire and from the Massachusetts marine
fisheries office — and one slightly older other.

The van, like most vans on research ships, is jammed
with instruments. A depth recorder, which traces a continuous profile of the bottom beneath the ship, hunkers
in one corner. A loran-C set, computers, monitors, radios,
files, workbenches jam the rest of the space. Four of us
fit, and the rest hang in the doorway.

Page gets out his charts, huge sheets marked with
loran lines intersecting in parallelograms an inch or so
across. On them he has drawn isobaths — lines connecting points of equal depth — in ten-meter gradations. The
Navy, Page says, already has detailed and relatively accurate maps of the American subsea, but they are classified
information. So he and other scientists whose work depends on accurate bathymetry have become their own
cartographers.

The charts, blue lines on grainy blue paper, show
Oceanographer Canyon, lying now some six hundred
meters below *Lulu*'s hull. Oceanographer slices further
into Georges than any of the dozen or so canyons that
were cut in the Bank possibly by glacial runoff and further developed by submarine currents and undersea ava-
lanches. Its head is cocked to the left like that of some
great worm, twenty-two kilometers inshore of the shelf
break, where the relatively flat continental shelf meets
the steeper pitch of the continental slope. Oceanographer's tail lies in 2,500 meters of water, 50 kilometers to
the south on the continental rise, the gentle declivity that
leads away from the foot of the slope to the abyssal plains
of the Atlantic. Joe calls it a classic canyon. There are
others out eastward that might match it, like Corsair
Canyon, but that is in waters claimed by Canada.

Unlike other incisions nearby, Oceanographer is still
an active canyon, eating away, eroding, slumping, carry-

ing on. And perhaps because of this — because of the strong currents generated by storms and tides and internal waves — the canyon is a rich place for some forms of life. Lobsters congregate around the rim and down in the gorge. They and the red and jonah crabs and the hake, redfish, and tilefish occupy burrows and holes and shallow depressions they or other animals have made. From worms to whiting, Oceanographer's inhabitants like their creature comforts; some of their excavations are so inspired that Joe and Page call them pueblo villages. Their efforts resuspend settled silts and clays and contribute to new erosion.

Oceanographer is no place to trawl: the terrain is too rough. One of the first geologists to study the canyons of Georges wrote that "if their features were visible, they would compare scientifically with the most impressive canyons in the world." Lobster pots do better than nets, though there are enough ghost traps in the canyon to demonstrate the risks in that line of fishing. Joe and a colleague took a bit of a risk of their own. Fisheries believes that its scientists should develop facts and not policy. Nonetheless, in a 1980 paper on Oceanographer that Page wrote with the two biologists, the authors recommended that the canyon be designated a marine sanctuary. That way, they wrote, "this kind of a natural laboratory could provide, in time, an ecosystem model . . . of inestimable value from which resource managers could more accurately assess the potential effects of human intervention such as fishing, petroleum exploration, marine mining or ocean dumping." This time, Fisheries listened — and put Oceanographer on its list of possibles. The past record of sanctuary proposals for Georges is not good, but one never knows.

We leave Page, charts spread out for the evening's bathymetry, and go, two by two, veterans and virgins, to be checked out in *Alvin*. The pilot doing the briefing is newly qualified, steady, thickset, and thickly freckled. By the time he has worked through the roster to the state fisheries man and me, he has almost lost his voice. The titanium sphere is seven feet across and encrusted with instruments. We, the future observers, learn to fold ourselves just so to avoid developing the painful chafing known as *Alvin* elbow as we peer out the side ports during the dive. We listen to the hoarse voice inches from our ears telling us what to do if the pilot gets a heart attack.

Your major goal, we hear, is to get to the surface. You come over here and hold these switches open, and they release your ascent weights. If you're still stuck, release your mechanical arm. Then the three main battery packs. If that doesn't work, turn this release on the bottom of the sphere. That separates the fore and after body and should free you.

"You're going to come up fast," says the pilot, "so be prepared for some weird oscillations. And when you hit the surface, we'll come and get you right off so you won't spend too much time upside down. You can still communicate with us through the underwater telephone, and if that doesn't work, you can always bang on the sphere."

In eighteen years, *Alvin* has had some close calls nosing around abyssal nooks and crannies or scouting for such booty as an inadvertently jettisoned hydrogen bomb over in Spanish waters. Early on, a sling broke on its launching system (hence the present cradle) and it came very close to taking its crew with it to the bottom. In the midseventies, it was entangled in a deep fissure on the mid-Atlantic ridge for the better part of an hour. But the sub is skillfully built and skillfully driven. And if something dire does happen, it appears we have someone on board *Lulu* who can help. He is with the Navy's deep submergence vehicle *Avalon*. But then it turns out *Avalon* can't mate with *Alvin* to allow survivors to pass through joined hatches to safety. That learned, we take something less than comfort from the motto on our failed savior's T-shirt: "If we can't get you, nobody can."

In the evening, *Lulu* begins the bathymetry. Page and Joe take the first watch. Joe sits by the depth recorder, pushing a button every minute. When he does, a line prints across the paper unrolling from the machine. Page reckons the ship's position on his charts, staring at the stuttering red readouts of the loran. Weeks and months from now, Page and his people at the Geological Survey offices in Woods Hole will be at work matching the time-marked depths with the tracks plotted on the charts, and Oceanographer will come a little clearer.

Most of the men of science sleep in two-man cabins in the after part of the van. The Massachusetts man and I are down with the crew in tiny two-man cubicles that run, inboard and out, most of the length of the starboard pontoon. At night the place is bloodred with night lights. It smells just a little like a cow barn, and that to me is a comfort.

The head is one of the cheeriest places aboard. It is kept sparkling by the crew. Every man dries the steel basins when he's through. There are no misfits here, just the normal complement of seagoing eccentrics, like the storklike crewman whose laugh belongs in a birdhouse. Almost all are young, here to be able to say they've done it.

Lulu bumps and grinds in heavy weather. When she does, there is not much else to do if you're off watch than stay in your berth. Some scientists have developed bedsores after a particularly lengthy blow. The crew has seen plenty of wind, for the ship has trundled from the hot vents on the floor of the Pacific to the rift valley of the mid-Atlantic ridge. Some of them spent only a couple of months ashore last year. They should be sick to death of each other. They don't seem to be.

The steward serves huge breakfasts to the youngsters in the mess over on the port pontoon. And then it's time for worship. We stand before *Alvin,* the crew swarms over it, in and out through the conning tower or sail, ad-

justing, checking, double checking. The sub's forward
port cover has a blue eye painted on it, and its white fiber-
glass skin and red sail hold the early sun. *Alvin* is so
squat it makes a lie of its twenty-five feet. It is a Polyne-
sian deity, fat and cradled. Joe and Page are taking the
first dive. They are both dressed in old blue clothes.
There is the slightest sense of the sacrificial in the way
they stand, silent, apart from us.

"Sub call," on the squawk box. "Sub call."

The ceremony begins. The engineer tests the cradle
and then lowers it between the pontoons until *Alvin*
floats free. Crewmen on each pontoon snub thick tag
lines to cleats and winches. The pilot, he of the lost voice,
clambers up a short gangplank and disappears down
through the sail into the sphere. Joe and Page follow him.
The sphere is sealed, and the chief pilot sets himself up
in the sail. The sub's huge propeller churns slowly inside
its white nozzle, and *Alvin* backs out from under. Two
swimmers dive from its tiny deck and unhook tag lines. A
Boston Whaler darts out from *Lulu* and takes off the chief
pilot, then returns for the swimmers. *Alvin* blows her
vents and settles. The sail disappears. The white hull
shows blue, bluer, and is gone. Dive one is in progress.

Alvin carries a transponder, a device designed to respond to a signal from the mother ship. Up in the pilot
house, we can hear the transponder ping every two seconds. The hydrophone operator beside me swings his
phones, lowered from *Lulu*'s bow, to get the sub's range
and bearing. *Alvin* comes in on the underwater telephone, the pilot's voice echoing as the acoustic signal
bounces. Nothing special, just some readouts from his
gauges.

In half an hour, the sub is on the bottom. I can picture
the two observers, squinched low, logging what they see
in the cones thrown by the external lights, recording
everything they can every way they can. Data is the busi-

ness at hand, very expensive data. "One, two, three cerianthans. A large silver hake in a pocket by a glacial erratic one meter across. Bottom is flat sand. Current: out of the north, about thirty centimeters a second. Time: ten-thirty-seven. Depth: five hundred and fifty meters." Talking into tape recorders, taking still color pictures, activating the recording device on the video machine. Thousands of bits of information fed year by year into computers to produce the sketchiest summary of the continuum of a submarine canyon.

Every fifteen minutes, *Alvin* gives *Lulu* its depth as it drives along. And does so again every time the sub's arm reaches into the chin basket on the bow and plunges a corer into Pleistocene clays as hard as cold butter or sweeps a sample bag across softer sediments. The hydrophone man figures his figures and calls his findings to the two-man science party on watch in the wheelhouse. The science watch writes the numbers down, adds the loran numbers showing *Lulu*'s location, and pencils a small cross on the chart, followed by a tiny stencil line of *Alvin*'s silhouette.

In the sub, it is chilly. The men will knock off in an hour to have lunch. It will be peanut butter and jelly sandwiches. It will always be peanut butter and jelly sandwiches. The chief pilot has so decreed. He claims it's the universal sandwich and that it doesn't get as soggy as others. The suspicion aboard is that the chief pilot has a thing about peanut butter and jelly sandwiches.

In the afternoon, *Penobscot Gulf* swings in for a visit. She is a rusty converted tug belonging to a man who fishes up and down the Atlantic coast, tying up often at Woods Hole and sometimes lending a hand to scientists there in need of a boat. He has been harpooning swordfish out here as they sun on the surface and he has one for us, he says.

"Thanks," *Lulu*'s captain says. "What can we do for you?"

"You've already done it." An offering to science. But to keep things right, *Lulu* sends back steaks and beer. *Penobscot Gulf* does not object.

"Sub call, sub call." Everyone grabs a work vest, a couple of webbing straps with flotation pads attached. The pilot house empties to give the captain and chief pilot peace in bringing *Alvin* home. The whaler is lowered and goes tearing off to the recovery site. *Alvin*'s red sail pops through the sparkle, and the swimmers tumble over and swim for it. The pilot has uncorked the hatch and is up in the sail, and *Lulu* tenders her cavernous rump to him.

Tag lines snake out and the swimmers attach them. "Bow lines," the pilot barks, "forward. Midships up. Midships forward." The linesmen run and slip on the pontoons, straining to keep the sub free and centered. That is crucial. Once the team failed, and in the resulting smash the great manipulator arm disjointed and sank. It was recovered months later, and that piece of bottom is now known as Alvin Canyon.

The cradle snugs up under the hull and steadies it. The pilot is out, and now Joe and Page. In a minute they are on deck watching their place of confinement rise to greet them. Both men look slightly shorter. Joe stretches and moans a little.

"Shark astern," calls the captain. There, where the swimmers were working, is a fin and, five feet behind it, the loll of a tail. A blue, someone says. Big, but probably not dangerous.

All during his years of diving, Joe has thought that on some lucky day he would look out his port and see Archie, the great squid, *Architeuthis*. A few have been seen dying on the surface or dead on the shore, meters and meters of tentacles and the thick body rotting. No sea serpent today. Instead, acres and acres of red crabs and sometimes magnificent lobster specimens. The dive track climbed the continental slope to the east of Oceanographer Canyon. Page says the current was sometimes pretty stiff, reaching fifty centimeters per second at the shelf break, about a knot.

The currents are tricky. Since the main flow around Georges is clockwise, sediment is often shunted southwesterly over the rim of the canyon. Tidal currents and internal waves tend to run up and down the axis of Oceanographer Canyon, sculpting large sand dunes on the floor and winnowing the fine silts and clay particles that presumably settle again in the deep sea. Page has a theory about the coarse gravel and boulders he and Joe saw today. He thinks they were probably rafted out here. Back fifteen thousand years ago, sea level was so low that these canyons were mere estuaries in Georges' flank. Ice rafts from the north, freighted with matter torn by the great rasp of the glacier from Canada, from upland New England, floated into the estuaries and grounded or sank as they melted and lost buoyancy. They left their template in fields of cobbles.

As the days roll, *Alvin* drives up the continental slope east and west of the canyon and makes four transects across it, from midpoint to head. At one point, the sub inches past blocks of slumped clay big as dump trucks, and the three men in their sweating steel bubble pray that the jumble above them holds as they creep along the foot of the wall. *Alvin*'s basket fills with samples: sediments, rocks covered with "potato chips" sponges unique to the eastern side of the canyon, and, one afternoon, a fourteen-pound lobster, its carapace only slightly crushed by the prosthetic at the end of the manipulator arm. Just right for a supper salad.

Each night, we do bathymetry. Now we have videotapes of the day's dives to look at. The illuminated floor drifts by in ripples and waves and flats. A hagfish there, surely one of God's ugliest attempts. Sea anemones. A large skate. Clouds of small shrimp and euphausiids.

Pueblo villages, the tenants peeking out. The biologists count more than forty species of fauna, and the top to a peanut butter jar. The chief pilot has struck again.

Joe is a flyer. He would have liked to have been an airline pilot. The two jobs, flying and diving, are similar, he thinks: 99 percent boredom, 1 percent terror. I never get to experience the 1 percent. Dive seven is scrubbed. It is a close call — the wind isn't bad, but a swell is coming in from the southwest and it is building — and that makes the cancellation that much more frustrating. Joe wants to wait, to see if he can get some return on his $14,000 for the day. We wait. The swells grow. Late in the morning, we head in, taking the beam sea. The scientists aren't moping. Six out of eight dives isn't bad in September, hurricane season on Georges. Joe and Page have built more knowledge of the canyon and started a comparative study of the continental slope just over the rims. And then there is that damned bathymetry, so boring, so absolutely necessary. How good is an explorer who can't map?

The chief pilot is talking to Joe about the time Walter Cronkite made a dive to a hot vent. I grind my jaw in envy. They were going down, the chief pilot says, when he lost rudder control. They made a dive of it, he says. "It came out on TV that I was concerned. Actually, I was scared shitless."

I feel better about my scrubbed dive.

The sun is hot and hard in the afternoon, and all the blues of the sea flow into black. The Massachusetts fisheries man sits on deck in a ruptured lawn chair studying old newspapers. He is a diver and a student of wrecks in his spare time, not so much for the salvage, he says, as for the find. Sweating and tanning, he says he found a wreck in Vineyard Sound, in fifty feet of water, the day before he came aboard. She was *Irene and Hilda,* the dragger missing in that late October storm of '80. Six men gone and, until now, no trace.

The Massachusetts man says he swam down and looked around. In the fo'c'sle was what looked like a bag. He reached into the bag, found it was a jumpsuit, found bone, and knew. "The head was gone," he says, shifting in the ruin of the chair. I remember the radio that night coming back from the storm. "Fishing vessel *Irene and Hilda*. Fishing vessel *Irene and Hilda*." Over and over. And, now, out.

6

The Federal Sea

THERE was once a herdsman who sought to better himself. "My herd is too small to bring me gain," he said.
So it was that he left the common lands of his village and
went away under the bright stars of morning. In the
mouth of the night, he returned with a new sheep.

"Hoo, hoo!" said the herdsman's brother. "I will not
believe but that I myself will go also." And he did that,
following the moon home with a ewe and her lamb. But
to make the long story short, each herdsman on the commons went away in his turn and added to his flock. The
grass of the commons grew thin, but each herdsman
said, "My sheep are not so fat as they once were, but
since I have more I am the richer." Until after seven
years and seven days the commons lay bare and trodden,
unable to support a cricket. Then the herdsmen beat
their breasts and went away they knew not where.

That is how they might tell the tale of the tragedy of
the commons in the West Highlands of Scotland. Garrett
Hardin, the ecologist who introduced his version of the
parable in the late sixties and saw it become part of the
environmental gospel of the seventies, was more to the
point. By tragedy, he said, he was referring not to unhap-
piness by itself but to what the philosopher Alfred North

Whitehead called the "solemnity of the remorseless working of things." Unlimited access to common property will work for centuries, Hardin wrote, "because tribal wars, poaching and disease keep the numbers of both men and beasts well below the carrying capacity of the land." But when society matures to the point of relative peace and order and good health, the tragedy begins. Each man is locked into a system that compels him to increase his herd without limit — in a world that is limited. "Ruin is the destination toward which all men rush, each pursuing his own best interest in a society that believes in the freedom of the commons. Freedom in a commons brings ruin to all."

What makes the tragedy tragic is that most of us, although we see it at work each day of our lives, reject it. We read of ranchers or mineral interests pressing the government to lease them more public land, even though it may be unsuitable for grazing or more suitable as park or wilderness. We read of heavy industry sending its toxic smokes into the ownerless air. Hardin is a Greek chorus. "The individual benefits," he laments, "from his ability to deny the truth, though society as a whole, of which he is part, suffers."

There are few commons left in farming country, and our agriculture, despite loss of topsoil and addiction to chemicals, is still a wonder. But there are no fences on Georges Bank. Hardin's law works well at sea, and few know that better than the fishery biologists who work for the National Marine Fisheries Service. They would prefer doing what they did when Spencer Baird first set them sailing and catching and dissecting on the old *Albatross*. But the oceanic commons is bare and trodden enough now so science must work in harness with management, in harness with Hardin.

Even in the early days, when the fisheries were the American promise, some saw truth in Hardin's view of the rush to ruin. "The most natural cause of it," said a colonist of a trough in the cycle of fish landings, "is probably the immoderate catching of them at all times of the year." Still, when politicians talked about protection, they almost always meant the industry and not the resource. Even after World War Two, the emphasis was anywhere but on fish stocks. Program after program came out of Washington: to help fishermen by putting their product into school lunches, by attracting young men to the draggers, by improving safety to cut down on disastrous insurance premiums. None of them worked particularly well.

What really turned things around was the Russians. Word got out in Washington that they were building fleets of huge factory ships to dominate world fisheries. The Soviets showed up with their giants at the beginning of the sixties, and then the Germans and the Poles and the Spaniards and the other splendid fishers that William W. Warner describes so well in *Distant Water: The Fate of the North Atlantic Fisherman* (Atlantic–Little, Brown, 1983). The new technology moved in floating cities, and it broke the back of the fisheries from Labrador to Georges and on southwest. At first the foreign fleets fished for species of little interest to New England ports, like squid and red hake. But in the middle sixties, there occurred one of those periodic blooms of haddock, or, more precisely, haddock surviving to a catchable size. There were more Russians on Georges by then than there were Americans. The foreigners swept the bottoms and the midwaters. To them, Georges was just one of many shoalings in the high seas, places with names like Whale Deep and Flemish Cap off Atlantic Canada, No Name Bank off Greenland, Bill Bailey's Bank off Iceland, Viking Bank and Tiddly off Scandinavia, Skolpen and Parson's Nose off the Soviet Union. Processors and trawlers moved along the chain pulse-fishing, setting their nets for one species until the catch became too small to bother with, and then moving on or setting their nets for another. The system worked so well that shortly it failed.

The intolerable had happened: the Russians were now the highliners. In 1960, just before the foreigners arrived on Georges, American boats were taking 90 percent of the harvest there, and most of the remainder went to the Canadians. Twelve years later the Yankees were taking 10 percent and the foreigners the rest. It wasn't just the long-distance fleets. New England's total catch had been dropping, with some surges here and there, ever since 1950. But you couldn't tell that to the captains. From Rhode Island to Maine, they sat in their small, wood-hulled, fifteen-year-old boats and roared.

The response from Washington was the Fisheries Conservation and Management Act of 1976, known as the Magnuson Act. At first, most fishermen thought it was the best thing since the diesel. The act extended United States fishery jurisdiction out to two hundred miles. We've kicked them Russians out, they said in the ports. Now there'll be plenty of American fish for American boats. Finest kind.

Nothing of the sort. The happy skippers either didn't bother to read the act or were engaging in the sort of denial Garrett Hardin found so curious. The Magnuson Act, which took effect in the spring of 1977, did limit foreign fishing to what Americans either didn't want or couldn't take. But it also declared that henceforth all fishes in the federal sea, with the exception of species like the highly migratory tuna, would be wards of the state, restored to abundance, protected against exploitation.

New England in particular should have known what that meant. For years, the United States had been trying to restore and protect the fish of the North Atlantic in company with the other countries fishing those waters. They called their organization ICNAF, for International Commission for the Northwest Atlantic Fisheries. The National Marine Fisheries Service was the lead agency in that endeavor, and its scientists and their foreign colleagues built such an imposing library of data that the North Atlantic is now the most thoroughly studied of any fishery in the world. But when data progressed to policy, ICNAF failed. Member countries either didn't enforce the toughest regulations or fought to keep them off the books. Enough did get on the books to rile New England fishermen. Not only were they operating under quotas and having to use nets of a certain mesh size and being denied big chunks of ocean during spawning season, it was their own National Marine Fisheries Service laying the law on them. And, they suspected, it was Fisheries scientists telling the foreigners where to trawl.

Washington pulled out of ICNAF shortly before the Magnuson Act took hold. And the new management? To handle the mare's nest of trouble, the skull-swelling problems of biomass and fishing effort, quota and catch that so bedeviled ICNAF, Congress chose to create eight regional councils made up in large part of amateurs. And who would advise the amateurs? Why, Fisheries, of course.

Bob Edwards is not your typical government scientist. He may have headed the Northeast Fisheries Center, a collection of laboratories with headquarters in Woods Hole. He may be a federal fisheries biologist. But he is not typical. He reads avidly, far from his field, and he studs his papers with quotes that have to give his more bureaucratic colleagues the hot twitch. A poem by Anne Sexton appears early in one report: "And man is eating the earth up / Like a candy bar / And not one of them can be left alone with the ocean / For it is known he will gulp it all down." Wagner comes along a little later; society, Bob says, is a bit like the opera *Tannhäuser*.

When I walked into Bob's office in the squat cinder building at the end of Water Street, he was playing with a toy tank. The thing was radio-controlled, and Bob was making it dodge around table legs and back out of corners. He pushed a button on the little console in his hand, and the tank went dead. "I think it will work," he said.

Something in his tone made me think I shouldn't ask about what kind of work he had in mind, not just yet. So I went to my purpose and asked him how things were in the fisheries science and management business. "That question always gives me trouble," Bob said, sitting down behind his desk. In his office, the visitor gets the view. The islands in the sound were bleak in the March sun and not a sail showed, not even the white slab of a ferry. "There are no overriding policies or objectives in our country that relate to the management of fish. Beyond the Magnuson Act, which says a few things like one shall not manage on the basis of economics alone, and one shall pay attention to what is vaguely defined as optimum yield, it is very difficult to say what our policy is."

Bob has a white, large, Scandinavian face. The room danced so with light from the sound that we both squinted at each other. "What are we managing for, really?" he asked. "Take the commercial versus the sports fishermen. Recreational fishing has quite a voice, you know. They want us to keep eight hundred thousand tons of mackerel out here so they can go out and fill their boats in half an hour. The fact that the bulk of those mackerel won't be used as food for the market is of no concern. Now, you get a commercial fisherman in there and he's going to want to maximize his economic situation. The minute he tries to do that, he's cutting across the goal of the recreational fisherman."

Fisheries biologists, Bob said, had gathered more than enough information to manage stocks optimally. "What's missing is any kind of a mechanism that allows for the transfer of that information to the groups that are in combat. And what's missing is policy. Without that, when you throw data into the arena of confrontation, you're in trouble."

This is clearly one of his pet points. "If you expect science, ever — in fisheries or atomic energy — to make a societal decision for you, you're crazy. We've tended to cop out, to say we can avoid all this muddling through, this emotional confrontation, by getting a hard scientific answer. But a hard scientific answer to what?" He raises both hands, palms up. "Venality? Esthetics? All of those are value-judgment words."

And the toy tank? Bob looked at it lovingly. He worked on woodchucks when he was a young ecologist, he said, and he thought he'd take it up again. "I'm too old to dig up burrows and see what the chucks are up to," he said. What he was up to with the tank was to design a radio-controlled, track-mounted videotape machine that could clank down the hole and do his peeping for him.

Linda Despres-Patanjo looks over her workplace. It is late April, and the southwest wind coming straight in from the sea carries relicts of winter. Linda wears a jacket fat with goose down. She is a little over thirty and married about a year — that's the Patanjo — and she likes and laughs about her life. She is the only female chief scientist at the Northeast Fisheries Center. "Woman is old," she says, "and I'm not a girl, so female."

Linda's place of business for the next eleven days, and for more than a third of any given year, is a deck. This one, surfaced with asphalt tile against slipping, belongs to *Delaware II*, a 155-foot, 18-year-old research trawler out of Sandy Hook, New Jersey. *Delaware* is owned by the National Oceanic and Atmospheric Administration, the parent of the National Marine Fisheries Service, and she is just about all open deck. The bridge bestrides it, so there is clear space from the great winches forward in the whaleback to the stern ramp sloping off into the water. The priority of *Delaware*'s design was room to handle and service a variety of experimental gear, but for some years now, she has fished but one net, the big Yankee-41. There are long, narrow laboratory spaces in the legs supporting the bridge, and booms and gallows to handle the net, and the rest is deck.

This is to be the third leg of the spring bottom trawl survey, part of a sampling of fin and shellfish along the Atlantic continental shelf from Cape Fear, south of Hatteras, on up to Nova Scotia. The federal government has been doing fisheries management surveys like this for the past twenty years or so. We're going out on the southeastern part of Georges Bank and then across the Northeast Channel, the front door to the Gulf of Maine, and on northeastward up to Brown's Bank.

The crew is aboard: the captain, making his first trip aboard *Delaware* in these waters, and the two mates, and the engineers, fishermen, and galley men. Science straggles on, most people walking from their offices in the Fisheries building. Men, mostly, a few women: a young black technician, an older white one named Evelyn Howe, who will record all that is recordable on her watch. Her watch chief and mine is John Nicolas, tall and bearded with a left arm that gives him trouble.

There is a cruise meeting in the mess. The cruise coordinator from Fisheries acts administrative. He tells us we will be working six-hour watches — six on, six off. Wear hard hats on deck. Keep out of the mess the hour before meals. Don't hop over the cables when the ship is working the net or the captain will turn around and come back and the offender will be put on report. The captain takes the other tack. Aw, you know what you're doing. Welcome aboard.

I find my bunk, one of four in the starboard cabin forward, and wrestle to get the sheets and blankets around the stiff mattress. Five of us — Linda, Evelyn, John, Tom McKenny (an amiable and deliberative downeaster from the Fisheries lab down in Sandy Hook, New Jersey), and myself — will be getting up at five-thirty in the morning, eating, working six to noon, eating, sleeping, eating, working six to midnight, sleeping, and getting up again. The schedule is different for Linda, who will steal sleep when she can: the chief scientist watches over both watches.

John and Evelyn show me our master. John picks at the nylon twine. "It's an old design and it's beat up. But we can't change the net. If we use a different one we'll get different results. We've got to fish the same way with the same gear, year after year." The net is set up to catch fish from a foot and a half to eight feet or so above the bottom. Its footrope moves on rollers. The headrope is buoyed with aluminum floats.

If you were to put one of the new remote-controlled television monitors on the bottom in the track of the net, first you might see cables passing to each side of you off in the murk. Then the doors — fifteen-hundred-pound ovals of oak rimmed with steel — running through the sand or bucking on the rocks, set like vanes to keep the net mouth grinning. If the monitor lens were wide-angle, you might see fish moving in to escape the plumes of sediment kicked up by the doors. Then wings of four-and-a-half-inch mesh to either side. Then the headrope, trailing a roof of netting. Then the monitor would be in the belly of the net, tumbling back along a narrowing tunnel of mesh to the tubular cod end. There, jammed in with a ton or so of cod and haddock and junk, slapped repeatedly on the bottom, the monitor would probably break. And, after a half-hour or so, be winched back and up *Delaware*'s ramp, swung over a wooden sorting or checker box, and, with the rest of the catch, released to crash down in a slither of fish.

The way *Delaware* does things, the monitor might not have much company in its net. The Fisheries sampling system is based on strata, regions of the shelf whose boundaries are determined by depth, water temperature, and other variables. Within the strata are mile-square blocks, and a computer is programmed to pick trawling stations at random from among these. The results, the analyses of catches huge, average, and piffling, are the stuff the statisticians ashore play with. As their data base thickens through the years, they improve the accuracy of their species-distribution models. They can't yet predict accurately how many fish are in the stocks; that is an incredibly difficult job, given the scales of space and time involved. But the analysts are getting good at working out where the fish are, often by knowing where they are not. "The commercial fishermen think we're crazy," says Evelyn, "fishing in the empty ocean. But when you have random sampling, you have to have parameters."

Delaware is under way, her stack trailing smells of burned and baking oil. We're in the lee of Martha's Vineyard, and the sea will be easy for a while. I am set by John to read a groundfish sex manual, a book of cold gonads to train my eye for the work ahead. Color photographs of open fish bellies show ovaries growing, the eggs in them turning grainy and then transparent, running from the vent. The ovaries shrink and change from spawning reds to resting blue-white. Testes balloon from narrow, crimped strips to bags tight with milt and then deflate to ragged, red-edged ribbons.

"Because most of the trained biologists in the lab are committed to preparing stock assessments," writes the manual's author, "they generally do not go to sea. The result has been that the surveys are often manned with persons who have little exposure to fish or anatomy." He describes how difficult it was for him to take pictures of fish fresh from the checker box. "Sometimes specimens reacted to the heat of the photofloods and back-flipped, . . . hakes had a particularly annoying habit of swinging their long fins across their gonads or the hand-written label." If you were split from pelvis to pap, O reader, might you too not flip?

The Fisheries folk ashore have made their wants

known, especially the stomach group, the biologists looking at feeding ecology. Among other things, they are interested in the oil strata, the tracts on Georges leased in Lease Sale Forty-two. Stomach contents can reveal whether the rigs are bothering the fish. This scientist wants sand launces frozen just so; that one, sea herring. Shawn McCafferty wants 125 yellowtails frozen. Margaret somebody is studying shell-boring algae and needs shells.

A note hangs on the lab bulkhead. "Please save, live, two-three dozen hermit crabs for feeding our octopus. Have a good trip. The Aquarium."

"Well," says Linda, sorting out forms and tally sheets, "we, the blood-and-guts biologists, have quite a trip ahead."

Orders are to study the spring distribution and relative abundance of demersal, or bottom-feeding, fish, especially cod, haddock, spiny dogfish, silver hake, and yellowtail flounder; to collect samples to be used in analyzing age and growth relationships, fecundity, maturity, and food habits; to check for disease; to take note of weather, water temperatures, and salinities; to catch fish eggs and larvae and other zooplankton in gauze-meshed Bongo nets. The stomach group wants its stomachs taken around the clock so it can measure variations in day and night feeding. "That's insane," says John. "We'll be sampling our asses off."

Evelyn shows me a stubby worktable along the gunwale just forward of the checker. Measuring boards are stacked underneath, each with a meter stick inlaid down the middle. Knives for gutting and shucking hang in a rack. The longer blades are for taking stomachs and scraping scales and going after otoliths, the small ovals of bone that help control balance. To get them, you slice back into the head from just above the eyes, lift the skull flap, and probe with a pair of tweezers. With scales and otoliths, you've got the age.

The Elizabeth Islands slip by to starboard. The twelve-to-six watch is stacking one- and two-bushel fish baskets by the checker, securing boxes of sample jars and XBT's — expendable bathythermographs, those bomb-like instruments that drop through the sea sending back data on the changing temperatures of the water layers.

Six fishermen are aboard, three to a watch. A few have put on some pounds. Their jobs can still be dangerous, but, without the commercial trawler's strain, the flat-out culling and gutting and icing and mending, the good food is apt to get to them. Here, they set out and haul back, but they are divorced from their catch.

The first haul goes out. A fisherman takes a hitch around the net with a line and another hitch around a winch wire. The net clatters across the asphalt tiles on its rollers. It slides down the stern ramp and begins to pull the heavy winch cable after it. Two fishermen stand on either side of the ramp, rigging the net pretty much the same way as the men on *Valkyrie* did. The net, swelling into its hunting horseshoe, and the cod end, with its brilliant chafing gear, all drop from sight with the flaring doors. The cables run out to the right scope, and *Delaware* settles down to fish. In ten minutes, the cables start to jump, slapping the deck. It's bad bottom, and the net is getting caught. The winches groan. We're hung up. *Delaware* backs down, and the net frees. Big Bill, the lead fisherman on the watch, hauls back. A ghost net, lost on the bottom but still fishing, slips over the lip of the ramp wrapped around one of our cables, a corrupted haddock jammed halfway through one mesh. There are all sorts of ghosts down there. Later we will bring up a couple of big offshore lobster pots, two of hundreds that will catch and kill lobsters until the wood rots or the plastic goes. And there is always the chance, tiny but tingling, that we may bring up some of the ordnance dumped or lost here or there along the coast. A boat caught something explosive off Virginia not too long ago, and whatever it was blew up and caused casualties.

The trawl has been too short for much of a catch. The cod end looks puny as it sways over the checker box. A fisherman pulls the release rope and it empties like a gut. The eyes. Rubbed copper in the eyes of flounder. Sea robins bloat, and their eyes are slits. Dogfish eyes glint green. Ocean pout whip on the bottom of the checker box, all lip and single-lobed tail. The net has a half-inch mesh liner that will catch just about everything except larvae; small lobsters; cunner; sand launces; a variety of flatfish — blackback, yellowtail, sand dab; long horn sculpin that buzz like the devil's vibrator when you touch them; skates; dogfish; sea herrings with backs the color of the noon sky.

Sorting comes first. You tell the developed female little skate from her big skate sister by the rough patches beside her vent. Male sharks and skates have claspers by theirs. Yellowtails have, praise be, just that — if they're along in their growth. Windowpane flounder are heavily spotted. Four-spot flounder have . . .

The six-to-twelve watch sorts, weighs baskets and buckets full of this and that, measures the requisite number of specimens, puts stomachs in jars and scales and otoliths in tiny manila envelopes. The net goes out again, and the cables jump again. But this time, when Big Bill hauls back, the net is gone. Seven thousand dollars, says John, plus a few more for the transducer mounted on the headrope to monitor the height of the mouth opening. "You're in luck," John says to me. "We only do this two or three times a year." The fishermen bind on a new net.

It is colder now that we are out on the shelf, cutting back and forth across the seaward edge of Georges Bank in spastic course changes to fit the random pattern of the trawling stations. Several of us wear long johns under our foul-weather gear. The night is double-dark beyond the work lights, and the stars spin above the black gallows.

Linda is always pleasant, almost always awake, and she draws us on. There are ten years at sea in what she does. When she started, working for the state of Maine, women weren't allowed on overnight cruises. Now she can look back on weeks aboard American and foreign research vessels. The Germans are best at fish-finding technology, she says; the French serve two-hour lunches with linen and wine; the Russians stay out for months.

John works with me at the checker box. My back and his mind, he says. A motorcycle accident is what gimped up his arm, and he has to maneuver it just so to get a purchase on the fish baskets. He used insurance money from the accident to help buy a house in his hometown on the Cape. He loves the house and his dog, a Dalmatian, and his girl, a printer. He does not love fish. John is a mammal man. Whales, he says. We'll see them soon. They're on the move northward up the coast. John has been going to sea for fifteen years. That, he says several times during the cruise, is enough.

A night haul brings up a lot of firm, pale sponges called monkey dung, and a great wolffish. He is dark gray, maybe thirty pounds, with broad, black vertical stripes and wizened eyes. His teeth are dark and dangerous. Into a basket he goes, by himself, and a crewman takes him after the measuring. The flesh is supposed to be fine, he says.

The weather worsens, and *Delaware* gets lively. Tom, the deliberate man from Maine, takes a slow roll across the lab in his chair. He props himself against the bulkhead but ten minutes later loses his balance again. John braces him with a big foot. "You're only allowed one of those per watch, Tom," he says, and Tom laughs a dry and modest downeast laugh. Evelyn is boiling a piece of sea cucumber in an old coffee pot to see if it really does taste like clam. It doesn't smell like it.

The smell of fish mucus on my gloves and slicker is worse than the sea cucumber. On watch again, I sweat and begin to pant as I help John lift a hundred pounds of dogfish to the scales. No hope. To the rail to let fly with great relief and some admiration at the muzzle velocity. Within an hour, I feel light and hungry. Things look interesting again, even the dogfish. A large cod thrusts himself up from the catch, a fish of forty pounds with a bronze back and olive spots. John cuts head meat from it — the cheeks and the vee of muscle under the lower jaw — and later fries some for us. It is nutty and chewy. But now all fish seems delicious; even the smell of my gloves makes me hungry.

The checker is empty, doused down with a seawater hose. The samples are logged and stored. There is time to cut fish for ourselves. Baskets of cod and haddock are lined up for the taking. George, one of the fishermen, smooths the edge of his springy filleting knife with a steel. His body is fifty and his face is fifteen. "Gee," he says, "I'm glad I don't have to do this for a living." He once did, in a fish-processing plant when he was a kid. He got so he could handle several hundred pounds of flounder in an hour. George is Nantucket born. His grandfather was lost at sea, he says. His dad dory-fished Georges Bank and once got separated from his ship. He rowed four days, eating raw fish. It rained so much he felt like a prune. When he was spotted off the Cape, horses snaked his boat through the surf and stampeded with him half a mile up the beach before they could be stopped. A Nantucket sleigh ride on dry land.

George has been a fisherman for thirty-five years. Now he wonders if he still is one, with the "chippin' and paintin' and aye-ayein'." People still in the business are making good money, he says. His friend Goofie Newfie told him he'd paid $16,000 in taxes last year. George talks and cuts. His stroke never pauses in the flesh. He grabs a haddock and turns the carcass with fingers sunk in the eye sockets for a good grip. In seconds the haddock is cartwheeling outboard, stripped of himself from just below the gills to the tail.

Someone starts talking about quotas and the council. He means the New England Fisheries Management Council, one of the eight regional groups created by the Magnuson Act. It has been having its troubles. There are some among the seventeen members who came in with considerable knowledge about fisheries management, or at least some part of it. But there are others who have had to learn almost from scratch, and do it in a goldfish bowl of public scrutiny. At the beginning, around 1977, the professional Fisheries advisers to the council were so worried about the condition of many stocks that they scared the council into accepting stiff regulations, most of them left over from old ICNAF days. There was scarcely a hook in the long line of regulations that didn't catch a gripe. Quotas were so tight that by-catches — hauls of fish other than the species being hunted — had to be thrown overboard. Limits intended to last a year were filled in six months.

Boats spent more time in port, and that gave fishermen time to attend council meetings waving placards that read "In Cod We Trust." When they couldn't make it to the meetings, their wives did. Many council members were processors or boat owners themselves, and it gave them the pip to sit in front of their seething fellows and try to pull things together. Skippers regularly broke the law. "Hailing" is the fisherman's term for declaring your catch on your return from sea. Haddock got hailed as pollock. Boats went from port to port, selling an illegally large catch in legal chunks. Since Fisheries and the Coast Guard didn't have the resources for more than spotty enforcement, even areas closed to protect spawning stocks were trawled frequently. When the enforcers began levying fines and confiscating gear, fights broke out on the docks between those punished and those whose luck still held. Up rose the cry of the Yankee restricted: "We were better off with the Russians."

In the remorseless working of things, though, the early optimism that fishing would be better with the Russians gone led to increases in the size of the fishing fleets, and these were making life even tougher for the council. Things got so out of hand that the Secretary of Commerce closed the cod and haddock grounds for two weeks one spring. When that didn't work, the secretary simply created a new calendar. The new quota year started in October instead of January. Finally, in desperation, the council scrapped the quotas for groundfish and switched emphasis to things like restrictions on the mesh size of nets.

George isn't exactly saying so, but it's clear he does not have exceeding fondness in his heart for the way Fisheries has dealt with his friends on the commercial boats. "Come to the point where fishermen got to take what there is to stay fishermen. Costs are going up — fuel, boats. Hell, the value of some *old* boats is shooting up, and the new ones — a lot of them go for over a million. To meet that mortgage, you gotta scrape bottom."

A crewman sides with Fisheries. Those council members can't tell a bell curve from a cod end, he says. When that big year class of haddock came in a while ago, they listened to the fishermen screaming that they were being denied their right to go after them. The members didn't pay much attention to the fishery biologists and their argument that one good year doesn't make a fishery, that you can't put pressure on a stock until it has a mixture of young and old fish.

"What fishermen don't understand," says John, cutting his own haddock, "is how much we've helped them. They think we don't know what we're doing, but we've developed new fisheries for them." Flip, splash. "Like offshore lobsters."

Silence.

"There's a lot of griping," says John. "It's somewhat like the farmers and the extension agent. We're like the agricultural extension agent. And those farmers — the successful ones cooperate, and the less successsful ones bitch."

George cuts fish, puts the fillets in plastic bags, and carries them off to his cache in the walk-in freezer by the galley.

The ocean is about to turn over. In some places, it already has. The XBT shows it's around six degrees Celsius at the surface and ten or eleven degrees on the bottom. That cold water should be sinking and mixing with the warm, bringing nutrients to the surface for the billions upon billions of eggs that will soon be floating around the Bank. But right now, on this day in late April, warm bottom temperatures bring bad luck. Near Oceanographer Canyon, the first haul of the night comes in all dogfish — trash. Tons of them, bulging the cod end. For over an hour, we sort and weigh and heave basket loads of bloody bodies over the side. We sample a few, cutting bellies to count eggs and pups — perfect sharklets attached to their yokes. Mist turns to rain as we throw away the last of them. The next haul is coming in. The bottom temperature at this station is lower. Maybe . . .

Tom stands by the worktable, cutting the few hake in the catch while Evelyn tallies. He is precise in laying the fish out for measuring on the centimeter scale. "Fahty, female, D." The hake, big-eyed streamlined fish popular in the European market, are a bit out of sync. The females are still developing and the males are spent. That could mean a bad year class. But young fish sometimes need time to work things out. Next year, probably, the hake will be more compatible.

The second haul is in, and scallops clatter into the checker box. Five bushels, one of the biggest hauls of them Linda can remember, and a sign of the times. In the old days, five bushels was nothing, even with nets like ours that fished over the bottom rather than on it.

A good way to shuck a scallop is to hold it in your hand, flat or white side up, hinge held in the base of your palm. Insert your small, humpbacked scallop knife forward of the hinge and cut back toward it, pressing up to separate the shell from the adductor muscle. Reach forward with the knife, pin the viscera to the top shell, and rip it off. There should be nothing left but round, sweet, white muscle attached to the shell in your palm. Slide your knife under it, and flip it into the bucket. Linda and I do this for two hours and get maybe fifteen pounds of meats, worth about ninety dollars retail. The scallops are healthy and young. They may be part of the bed the New Bedford scallopers have just found. If they are, they won't last long with the boat mortgages and all.

I go to sleep at two in the morning after the shucking, listening to the snoring sound the bow makes as it cuts the swells. The sea is down. It's a long way from dawn, when a fisherman on our watch — Joe, I think — looks out on the steep slides of water and says, only half for the laugh, "God, please make this thing flat."

We run on out to the northeast part of Georges and across Northeast Channel to Brown's Bank, boneyard for nets. Here bottoms are sharp and rough and apt to be covered with trees — soft but stubborn branching coral. We tear out bellies and strip wings and skip a station when the bottom gets too bad. Blackfish, or pilot whales, run by us, and dolphins ride under the bow. Birds come to a ghost dinner bell. The minute the net surfaces, there they are a-running down an empty sky.

Lobster pot markers are everywhere. We haul in big lobsters, six to eight pounds, and they go to the live well for suppers ashore. The other watch gets them. Ours nets some more scallops and then a halibut, a four-foot flatfish strong as spring steel. Swede, the other fisherman on our watch, steaks it, and everybody around gets a piece. And again, that's different, a delicacy. Back in the brightest days of the fishery, halibut was just halibut, and haddock was a trash fish.

The traffic picks up on Brown's Bank. These are Canadian waters. The boats are from Yarmouth and Lunenburg and other Nova Scotian ports. We swerve to miss the cables behind a blistered old scalloper.

There is a goosefish in the late afternoon haul, almost a meter long, sombre and ugly. John has her on her back, her pelvic fins like a baby's feet on her flaccid, gray-white belly. John sexes her, pulling out yards of egg veil, a yellow sea scarf. He turns the fish over and slices its huge head in half vertically. He gets one otolith and loses the other in the welter of meats and fluids. The split mouth gapes, great nails of brown teeth angle back. An eye opens, a glint of emerald in the late sun.

The port winch on *Delaware II* has been creeping for a couple of days, giving the fishermen fits. If the cable moves when you're shackling or unshackling those doors, you could be messed up. Suddenly, the brake goes altogether, with three hundred fathoms of wire out. The captain comes down for a look. They'll have to winch it back end for end, using the starboard winch.

We're dead in the water. A Canadian government fisheries vessel comes scurrying in for a look. We saw you making ten knots and then three and then nothing, they say. Just checking. When a boat out here isn't moving, it could be in trouble, or it could be waiting for a drug drop. Just to be sure, or just for practice, a Canadian Coast Guard patrol plane rigs us, buzzing down the starboard side. It's a busted trip. We'll have to go back to Woods Hole to repair the winch. We were set to bitch for two more days about being out here, working up to a fine return: the sailors home from the sea with the weekend right there for us. Now we're early and surly about it. But then John decides to retire his foul-weather gear. You can't see the yellow of it for the fish gurry, and the pants are ripped. We pipe it over the side with an impromptu

Fisheries salute. The right hand waggles at the forehead
in poor imitation of a knife after an otolith. The left makes
slicing motions up and down the belly. Then we head
down to the mess. Fifteen minutes to the next seating.
Fifteen minutes to gobble golden haddock and beans and
potatoes and a light and lovely cake.

Bob Edwards is no longer head of the Northeast Fisheries Center. His job now, in the spring of 1983, is to
advise the Department of State on how best to build a
strong biological argument before the World Court for
U.S. ownership of Georges Bank rather than a split jurisdiction with Canada. The dispute is taking on massive
momentum. In Washington and in Ottawa, international
lawyers, economists, oceanographers — squads of specialists — are working against lists of deadlines for this
position paper, that report. In the end, the briefs and
their addenda will stand taller than Bob himself, taller
than six feet. The walls of Bob's new office in the Fisheries building in Woods Hole are papered with charts of
fish distribution, population analyses, and the like. Semi-
secret, he calls them.

The walls of Bob's old director's office are being lined
with managerial panelling. No more cinder blocks. But
the view of the sound is the same. The remodelling is appropriate. Bob's successor is not a working scientist,
though he started out in biology. He was the National
Marine Fisheries Service's regional director for New
England, managing programs affecting scallops and scallopers, groundfish and trawlers, until his superiors in
Washington asked him to take over the research end of
things. His name is Allen Peterson, no relation to the
Susan of the same name who plies her anthropological
trade up at the other end of Water Street.

The day before I came to see him, Allen Peterson had

given a talk to Susan Peterson and a group of other scientists from around Woods Hole on what we should be managing fish for. Allen didn't want to repeat to me what he had said, because he didn't want the public to take his private ruminations for the official Fisheries line. But as we went along, pausing when the panellers' hammers got too insistent, it was clear that he was thinking about what happens when you mix problems having to do with fish and problems having to do with people.

Take Maine, he said. "The sardine industry there is part of the social culture, a form of livelihood. But I think biologically the harvesting of sardines [young herring] is not the best way of managing the herring fishery. Right now, there's a big complaint down in Maine. The sardines didn't come into the weirs, so the canneries are having a lousy pack and people aren't getting their wages. The sardines are sitting offshore, but purse seiners aren't allowed to come in and catch them. The people in Maine are fighting over fishing techniques rather than saying, 'What we really want to do is keep the canneries operating.' "

The hammers hammer.

Take scallops, Allen said. "We said we wouldn't be concerned about the pounding scallops are taking because economics will take care of things." Economics hasn't. Scallops are now over six dollars a pound and Allen thinks they might go to ten. "The scarcer they get, the higher the price, the more people want to catch them. The more you get illegal activities." The same thing, he thinks, may happen to Atlantic salmon.

Allen is an obversion of Bob Edwards — dark-complexioned, compact, reined in. But the memory of the talk he gave on his management ideas clearly excited him. "I said I don't think you can deal with scallops as an individual species, or codfish either. I think you have to treat all fishes as one management unit, so you can move fishing effort from one species to another. In Canada, there's a fight between inshore and offshore. Here, you can go from one to another, if you've got the money to buy the gear. There you can't. It's limited entry."

More hammering.

"We and the Canadians are going to Georges Bank right now and taking, you know, tomorrow's proceeds," Allen said. At present, legal scallop catches must average at most thirty-five meats to the pound. It started at forty meats and was supposed to drop to thirty, but Allen, in his previous job, agreed to a compromise when the scallopers told him they'd go broke trying to find meats big enough to make the count. The boats are still mixing in a great many young scallops with a few older ones. The Canadians do it too. "If you stop 'em from doin' it, you put 'em out of business," Allen said, his mid-Massachusetts accent coming out in the drive to make his point.

"From the fisherman's point of view, it's 'Hey! lemme stay as long as I can, and I'll cross my fingers and make a prayer that something will happen in the meantime.' To me as a manager, the choice is puttin' 'em out of business and having scallops for tomorrow. But if you stop the fishery — and that's the only alternative right now — that throws you into a whole range of economic consequences, and now you're into social management. And you can't put that on the back of fisheries management.

"Right now," Allen said, "we're doing a terrible job with social management. I joke a bit about it, but it's pretty hard to go broke in the fishing industry. In fact, if everybody who told us they were going broke went broke, we wouldn't have any problems. Fishing just isn't a capitalistic activity. It's like small farming." Allen grew up on a farm. "You do it because it's in the blood. And you don't worry about things like depreciation. What's going to happen is that our fleet is going to get old, and there won't be any capital to replace it. We're going to end up like the steel industry, no longer competitive."

The policy here as everywhere has been to provide

support for the people involved rather than solve ecological problems. It doesn't go much further than the idea that it is important to have a fishing industry on Georges Bank. "As long as you keep that up," Allen said, "you don't winnow out any of these fishermen. You keep increasing the fishing effort. You keep aggravating the problem." That must have been some talk he gave to the scientists.

Allen opened his office door for me and another door in his mind. "I won't speculate if what we're doing with oil rights is the best way to handle them," he said. "There, the government is giving people the exclusive rights to utilize the resource, but it retains the ownership. My view is that we could do similar things in the fisheries. We might be able to control access by granting certain kinds of ownership for certain periods of time. That way, you might get away from the tragedy of the commons."

He laughed. He said he didn't know how most of the oceanographers in his audience took to his ruminations the other day, but the social scientists present had perked up. His nonrelative Susan Peterson was coming down the street in a few minutes from the Oceanographic to hear more.

7

All Rise

THERE was a wind and it blew. A hard wind, sweeping over the coal seams and the timber of the West, over the proven and unproven oil bottoms of the American offshore. James Gaius Watt was now Secretary of the Interior, possibly the most controversial secretary in the history of the department, and he was blowing hard enough to crack his cheeks. James Watt believes in commitment — mainly to the Judeo-Christian principles he found when he responded to the altar call of a Pro-Gospel Businessman's Fellowship meeting. He believes in battle, and during his two and a half years at Interior, that meant battle against what he called "commercial environmentalists." They are, as he told James Conaway in just about the only interview that revealed something of the man himself, "hard-core, left-wing radicals manipulating the press. . . . They have a conspiracy of shared values. Their real objective is partisan politics to change the form of government."

It was Watt's conviction that "the most dangerous way to bring oil to American shores is to put foreign crude in foreign tankers and point them toward the United States." Domestic production had to be increased radically, Watt said, since his predecessors had leased only

2.5 percent of the outer continental shelves in twenty-eight years, and most of that had been in producing areas; 19,000 of the 20,000 offshore holes had been drilled off places like Texas and Louisiana, and only seventy holes in frontier areas.

Watt's idea was to open public lands so that, as he put it, inventory could be taken of their resources. For the offshore, he announced a plan that would offer almost all of the federal sea bottoms in forty-one lease sales over five years — about twenty times the acreage his predecessor, Cecil Andrus, had had in mind. Watt said he was going to streamline procedures to halve the time spent in the leasing process and to assure that the market, not the government, would select the lease tracts. No need to worry about serious environmental damage. The industry had a fine safety record. "In the last fifteen years," Watt said, "there have been no oil spills of lasting impact that can be based on drilling." Besides, there was a quarter of a billion dollars' worth of offshore research to draw on if it proved necessary to build in more safeguards. Watt said his program would "enhance the national security, provide jobs and protect the environment while making America less dependent on foreign oil sources."

Frank Basile, the Bureau of Land Management official who handled Lease Sale Forty-two that cold December day back in 1979, did what he could to put out the fires that Watt started in New England. His new employer, Frank said, wasn't thinking of actually leasing every acre of the offshore. No, no. What Secretary Watt wanted to do, Basile explained, was to "make sure that every sale includes the best acreage available."

A new office would handle the marathon. Watt called it the Minerals Management Service, and he built it, on his own, out of the Bureau of Land Management, some functions of the Geological Survey, and almost everything else Interior ran in direct support of offshore activities: scheduling, leasing, collection of royalties. The head

of the new office's offshore division told me the objective
was an attempt to develop large fields. Once those were
producing, it would be a lot easier to bring in smaller
fields nearby and plug them into the storage and trans-
portation systems built for the giants. "In the past," he
said, "we haven't allowed industry to look for their own
best spots." He hoped the companies would look early
and often. "We'll go back to the areas again and again."
He was talking about eighteen huge planning areas Watt
had set up from Maine around to Alaska. The North Atlantic planning area runs to fifty-seven million acres and
is just about average in size.

I asked Cecil Andrus in the early fall of 1982 about
Watt's accelerated leasing plan. "It's no different than
the rest of his rhetoric," he said. "He talks a big story and
thrashes about, and he hasn't accomplished one thing.
The oil companies have been saying, 'Look! We don't
have the money or the equipment to bid on that much
land.' " Inventory of national resources did sound like a
good idea to Andrus, but the best way to get at it would be
to "go to Congress and get an appropriation so the federal
government did the drilling. But Watt and his people are
just putting it out to bid. Do you think that if a company
uses its money to go out there and explore, and if it finds
something, it won't want to extract it?" A pause. An explosive "Hell, no!"

A few days later, I asked the same question of Douglas
Foy, the head of the Conservation Law Foundation and
Andrus' chief opponent in the legal battle over leasing
Georges Bank. "Basically," he said, "Watt's just using
blue smoke and mirrors. If he wanted inventorying, he
would be issuing exploratory leases with options to purchase development rights depending on what's found."
When Cecil Andrus was secretary, Doug said, it was hard
to go into court and argue that Interior was making irresponsible decisions and sacrificing living resources for
mineral resources, "because they would turn to you and

ask, 'Isn't Andrus supposed to be a really fine Secretary?' And the answer was yes." It would be different now. "I think James Watt is the best thing to happen to the environmental movement in five or ten years," Foy said sardonically.

In public, at least, Watt seemed to bask in the light of his own brush fires. He paid no attention to those who said he was moving too fast; or those who said that he would run into so much litigation from the environmentalists that the oil companies would back off; or those who said that so many tracts on the market would mean depressed bidding and thus lower revenues for the federal government; or those who said that the supply of oil and gas under the continental shelves simply wasn't sufficient to reduce the nation's foreign oil dependence as dramatically as the secretary insisted it would.

Watt's characteristic response was to return to the attack. He began submitting names of his advisers to Republican party scanners, who quickly spotted potentially troublesome Democrats. One was John Teal of Woods Hole, who got summarily dropped from one of the secretary's scientific counselling bodies. Then Watt went after Congress. He wrote a bald and blistering letter to the chairman of a particularly critical House subcommittee: "I cannot understand or interpret your motives. . . . It is much easier to explain to the American people why we have oil rigs off our coasts than it would be to explain to the mothers and fathers of this land why their sons are fighting on the sands of the Middle East, as might be required if the policies of our critics were to be pursued."

Litigation, Watt was fond of saying, bothered him not at all. "We are gearing up," he said upon approving his accelerated leasing plan early in 1982, "with the expectation that there will be litigation everywhere . . . and it will be brought by those who don't want to create jobs . . . or improve America's economy." Watt's expectation, if not his description, of the opposition was accurate. One of Watt's first moves offshore, in February of 1981, was to announce he was moving ahead with some leasing off northern California that the Carter administration had put on hold after having lost a suit over the matter. California took Watt to court, arguing that his intention was in violation of the Coastal Zone Management Act of 1972, which says that federal actions must be consistent with plans for coastal management developed under the law by coastal states and approved by the Department of Commerce. California won against Watt as it had won against Andrus. Nine states joined in that suit, among them Massachusetts. Rehearsal, you might call it.

The head of the Massachusetts coastal zone management program is a slight and sharp regional planner named Richard Delaney. It was Delaney's job in 1982 to oversee a group of economists and oceanographers and others hired to implement the objectives of the state's coastal plan, one of the first in the nation to be approved by the federal government. The issue fascinated him, the twisting and turning as Watt first refused to issue what is known as a consistency determination in the California case (a state can demand that a federal agency publicly assert that its proposed actions are consistent with that state's coastal zone management plan) and then tried to change the federal law to exempt offshore leasing from consistency requirements. Massachusetts' own Coastal Zone Management Office lent a hand in the preparation of California's consistency case against Watt and followed the suit as it wound up through the system on its way to the Supreme Court. "The issue is jurisdictional," Delaney told me in the fall of 1983. "Massachusetts and California and other states and environmental groups argue that affected state governments do have the right to review a lease sale. The Justice Department, arguing on behalf of Interior, says a lease sale is not an activity that will have a direct effect on the coastline. We say it sets in motion a chain of events that has an inevitable impact on our coasts and our fisheries."

Delaney and his colleagues in the Massachusetts state government weren't thinking solely of jurisdictional principles when they went to the aid of California. They knew what Secretary Watt had in mind for Georges Bank: Lease Sale Number Fifty-two. Work on Fifty-two had actually begun under Cecil Andrus. The Department of the Interior made its tract selection for Sale Fifty-two while the successful bidders in Sale Forty-two were still working their way through the bureaucratic puzzle game of getting the proper permits to drill. The department held hearings on the draft environmental impact statement for the sale just a couple of weeks before *Rowan Midland* spudded in on Block 312 late in 1981 and began making hole for Mobil. Fifty-two covered a lot more acreage on and off Georges Bank than Forty-two had: 2.8 million acres against about 660,000 — and many more tracts: 488 as against 116. Almost half the tracts were in water over 3,000 feet deep, within the industry's ability to drill but not yet within its ability to produce.

The new impact statement was a chapbook compared to the one Frank Basile had put together for Forty-two — only about five hundred pages all told, including the appendices. "You get better at things as you go over them again," says Basile. "But it's still a bit of a struggle to satisfy everyone within the space limitations we adopted." More than a bit. The new, improved shorty drew about as much fire as the old almanac: it was hard to follow as it bounced from alternative scenario to alternative scenario. Environmentalist critics doubted whether the biological task force, the compromise worked out for Lease Sale Forty-two, would also oversee Fifty-two. The second sale came too soon, they said; the results of the studies the Bureau of Land Management had funded to examine the effects of drilling in progress should be thoroughly

studied before opening any more bids. What environmental risks would be run learning how to produce at
those vast depths, they asked, and why did the oil industry have to educate itself on deep-water operations in a
piece of ocean abutting a world-class fishery?

Interior, having scheduled Lease Sale Fifty-two for the
late summer of 1982, put it off to the fall and finally to
March of 1983. Richard Delaney spent a good part of that
grace period on airplanes to and from Washington, by
himself and with others from the commonwealth (even
Governor King was involved) trying to get Watt's people
to delete sensitive tracts, ninety-eight in all, in order to
make the sale consistent with Policy Number Nine of the
Massachusetts coastal zone program. Number Nine simply says that Delaney's office has a mandate to accommodate offshore oil and gas development "while minimizing
impacts on the marine environment, especially on fisheries, water quality and wildlife." To Delaney, that meant
tracts on good fishing or nursery bottoms had to be taken
out.

What Interior did at first was to stare through the peri-
patetic Delaney and declare that as far as it was concerned, Lease Sale Fifty-two was consistent with Policy
Nine and everything else he had in his bag. Delaney held
a public hearing to get some extra ammunition for his
reply. Even before the hearing, he was willing to say
publicly that if the standoff continued, "a full or partial
injunction would have to be sought by the state to prevent the sale."

Several dozen people showed up at Delaney's hearing
in December of 1982. One of the country's most serious
students of whales was there, and he was worried about
what might happen to his subjects. "The first shot the
Pilgrims fired was to harass a right whale," he said.

Angela Sanfilippo was there, and her friend and fellow
fighter from the Gloucester Fishermen's Wives Association, Lena Novallo. "The state of Massachusetts should

feel proud to provide food fish," Angela said. "Every day our fish fly as far as California. Wisconsin has cheese, Idaho has potatoes. This state has fish product. The people in Washington should know what food fish are all about." She would try to get more information from the fishermen for Delaney to use. She would try to make some charts "to show where the fish have completely disappeared." Angela was not waiting for the professional monitoring studies of drilling on Georges to be completed. She believed what she heard from the docks. "The fishermen see the circles in the water around the rigs caused by the drilling muds. The ocean water is not the same color any more. It's affecting our fish."

The two women worked beautifully as a team, Angela angelic and Lena bold as brass. Lena is square, stocky, her voice a natural loud-hailer. You would want to avoid serious confrontation with her as you would with a fishwife guarding her stall. "I have four generations of fishing in me," she said. "My grandfather, my father, my husband, my son. Our fishermen have an industry providing food for the world — food we need. When Almighty God created the world," Lena said, "He made one Georges Bank and He blessed us with it, and it's up to us to protect it." At one meeting, she said, she had met an oil man. "He had one of those big hats on, so I knew he was from the oil place. He said, 'Why do you think oil and fish don't go together?' I said, 'They do go together.' He said, 'They do, but you didn't testify to that.' I says, 'Wait a minute: olive oil, vegetable oil and peanut oil, not crude oil.' "

Lena, Angela, Richard Delaney, everyone at the hearing knew that things were not going well for the leaseholders out on Georges Bank. The oil companies had drilled eight holes and every last one was a duster — dry, not enough oil or gas even to consider producing. Not much of a return on an investment, counting bonus bids and drilling costs, of about one billion dollars. *Rowan Midland* and the three other rigs working the Bank had all gone by the fall of 1982, *Midland* leaving behind an anchor on the bottom, to be dealt with by a cleanup outfit, and two plugged wells — the first, on Block 312, costing a budget-gutting $36 million. The rigs had been hit by another spell of awful weather after I left *Midland* — an April howler that got one rig super so upset he called for the Coast Guard. No rescue could have been attempted in that screaming wind. Fortunately, the weather eased, and none was necessary.

A few weeks after the rigs left Georges, I went down to New York to see Bob Graves, the Mobil vice-president in charge of hunting oil. Bob looked a little tired, big lids a little further down over his eyes than when we talked before Mobil started drilling on 312 — but given the global scope of his job, it was unlikely that two dry holes on the New England frontier had contributed much to that. Bob was, as usual, in his shirt-sleeves. He pushed a map showing Mobil's acquisitions on Georges across his desk and put a big finger down on it. Mobil had drilled on two plays, or prospects, he said, the reef and some structures — likely oil traps — to the east of it. "Our big concern was whether we were going to find any reservoir rock." They didn't drill through any of those porous formations essential to a decent find, at least none to speak of optimistically.

"Sometimes reefs can be extremely tricky," Bob said. "We've seen wells where there was no porosity and then, less than a mile away, we've seen just the opposite." *Midland*'s hole at 312 was tight — nonporous — in the sections that had otherwise looked promising. No hiding place down there. *Midland*'s second hole, on the structural play at Block 273, showed thin streaks of porosity — and nothing of value in them.

Bob thought there was perhaps one long shot left to drill on the structural play. As for the reef, Mobil probably would want to trade its data on the rocks found at 312 for lithologic information from Shell, which had drilled in a block just to the southeast. "If there is a difference between the two holes," Bob said, "that might mean we could find some porosity elsewhere on the reef. But if the data is identical" — a weary chuckle — "another balloon has burst on us."

I asked Bob about the deep-water leasing scheduled in Lease Sale Fifty-two. I had heard talk about a formation of supposed promise called the Great Yellow Reef that sneaked up the Atlantic coast way out there. "I know," Bob said. "Billions of barrels and all that stuff. That's what the reports say in the press. But I've got to tell you, that is an extremely high-risk play. A lot of rocks are Tertiary, pretty young. There is a lot of unconsolidated sediment and that doesn't make an effective seal to trap oil."

The cost of drilling in deep water and the questionable oiliness of the barrier reef buried beneath it would signal caution to the prospective leaseholder under any circumstances. At the time Bob and I were studying the markings on his map, in the late fall of 1982, oil was no longer the dangerously scarce item it had been perceived to be just a few years before. Recession, the extraordinary effects of oil conservation by consumers, these and other global swirls and countercurrents had produced an oil surfeit, a glut. The oil companies that imported a lot of OPEC crude — Mobil high among them — were caught by their contracts. The Saudis and their colleagues were charging thirty-four dollars a barrel at the beginning of the year. "We couldn't recover that in the marketplace," Bob said. "There was a loss of up to five dollars a barrel. Multiply that by fourteen or fifteen million barrels a day — tremendous loss for the companies." Things were somewhat better now, he said, but most oil economists still saw flat or falling prices out in front of them for a while. "This put the fear of the Lord into everyone. Once you start squeezing profits, there is no way you can continue spending like a drunken sailor." So the demand for drilling was down. Use of drilling rigs had shrunk 40 percent since the peak in 1981. Costs of leases had dropped. "Supply and demand," Bob said.

The drilling results on Georges would have given Mobil pause even if everything else were rosy, Bob said, walking back around the big desk to his swivel chair. Given the state of the oil market, Mobil would have to slow down and look at the data very carefully. "We don't have too many darts left to throw up there. Those that we do throw, we gotta make damn sure we're putting them in the right place."

I got up to go, and the goodbye talk got around to oil and fish. "When you go into a new area like the East Coast," Bob said, "there's hardly any amount of p.r. that you can do that will alleviate the kind of problems we ran into on Georges Bank. Because these are critical problems. Did you read the Greenpeace article the other day about prohibiting any further drilling there?"

I hadn't, but I remembered Greenpeacers in their rubber rafts waving placards at *Zapata Saratoga.*

"Well, you know, that's the kind of thing you're up against. But there is absolutely no question in my mind that the fishing industry and the oil industry can coexist. I mean, hell, the platforms are a benefit to the fisheries. There is no question about it. Those platforms are artificial reefs. They collect fish."

Bob's secretary was waiting with a look that said terminate transmission. "I'm a fisherman myself," Bob said. "Fished a lot in the Gulf. And off Montauk and Block Island. I don't want to see any fisheries ruined."

Bob shook my hand and began reading a phone-message slip his secretary had just handed him. I took the panelled elevator down to the Mobil lobby and out past the security guards. By the time I was out on the street, I remembered some earlier comments about platforms as artificial reefs. They came from John Teal, the Woods Hole biologist-chemist-ecologist. "Platforms do attract fish," Teal had told me. "People can hang a line down by a platform and catch fish they can't catch otherwise, which is fine if you happen to be a fisherman who likes to sit in a small boat and catch things. But what about the fish not inclined to be associated with platforms — the fish of the fishery?"

Teal had talked about developing an experiment to answer his question. If oil were found under Georges, he said, a few production platforms should be set up out there as controls — platforms just sitting, not making hole or pumping oil. What happened to fish near them could be compared to what happened to fish near the production units. "That would be very nice," John said, "and it isn't going to happen."

Richard Delaney and the commonwealth kept after the tracts they wanted deleted from Lease Sale Fifty-two. In January, when Michael Dukakis and Edward King changed places again in the State House, the new Governor of Massachusetts maintained the pressure. "It really wasn't until the eleventh hour," Delaney says, "that Interior came up here to talk." But not about deletions. "What the department had in mind was new language in the lease stipulations to protect the fishery."

The two sides sat down to bargain, and Interior gave a little. Steve Leonard, one of Attorney General Bellotti's lieutenants, the one responsible for environmental issues, says, "They wouldn't give us everything we wanted, but they came close, and we came close to taking it." But again, the Conservation Law Foundation made the difference. "They put countervailing pressure on us not to take it, and we didn't," Leonard says. With Dukakis in, the Conservation Law Foundation had more influence. "They see themselves as part of the Dukakis constituency, or representing it."

Douglas Foy thinks the commonwealth had been negotiating over Georges for so long that it had caught a case of infectious accommodation. "What happens in the dynamics of negotiation is that eventually everybody starts wanting to have results. You tend to forget what your bottom line was."

Doug and leaders of other environmental groups sat down with state officials, he says, one afternoon early in 1983 just before the decision was to be made on what Massachusetts should do about Sale Fifty-two. "We sort of beat on the table and said, 'This is irresponsible. You state people aren't sticking to what you told everyone you were going to do: hold out for those deletions.' To his eternal credit, the new state Secretary of Environmental Affairs decided that he was going to overrule his staff and go to the mat."

A few weeks after that beating on the table, some of Rich Delaney's coastal zone management people walked up the short distance over Beacon Hill and compared notes with Foy's lawyers. They were at it late one afternoon, late enough for Doug to emerge from his pupal indoor stage. *Whsssk!* Gone the suit and tie. *Whsssk!* On with racing shorts and shirt and cleated shoes. As the conferees ran their courthouse scenarios, the outdoorsman, freed of his cocoon, mounted his bicycle and pedalled away westward in the biting air.

What did I tell you, ladies and gentlemen! A real rouser, an even match of will and wit. Round one to the indomitable Mr. Foy and the commonwealth of Massachusetts. Round two to the Honorable Cecil Andrus and the federal government of these United States of America. And now, ladies and gentlemen, in one minute, round three of the great water fight, the battle for Georges Bank!

It is bitter, too, in Boston's Post Office Square, outside the building where federal courts for the region sit. It is Saturday, March 26, 1983, and the guards in the lobby are a little edgy about letting the first visitors go up the elevators to the courtroom on the twelfth floor. Visitors? On a weekend? In a while, a U.S. marshal arrives to man the security check-in desk. The courtroom is cool and blue. The dais of judgment sits at a judicial height up a cliff of blond wood, backed by thirteen gold stars on a black background. One of the three chairs on the dais will be used today, by Judge A. David Mazzone of the Federal District Court.

Doug Foy comes in with his team, two or three people. He is wearing a blue-gray pinstripe, a blue shirt, and a large yellow tie. He saunters over to the lawyers for the oil companies to talk about who needs what time for argument and oratory. The same with the pod of government lawyers, led by a trim and serious woman named Margaret Strand.

"All rise."

All rise, the spectators struggling up out of their narrow pews. Judge Mazzone, tall and long of face, asks some of the crowd to move over to the jury box. We do, sitting thankfully in the big blue leather chairs. Someone says we'd really be having a space problem if it weren't for the fact that many of the local environmentalists are at a conference out of town.

Foy sits with Steve Leonard. The Justice Department people cluster around Margaret Strand, sitting straight and serious a few feet away from the plaintiffs — or, to be more accurate, the plaintiffs present before the court: Foy and Leonard have been joined by Greenpeace and Massachusetts Audubon and Angela Sanfilippo's Gloucester Fishermen's Wives Association and others.

The oil-company attorneys, sombrely besuited, take up a position off the government's port beam. Maneuvering starts immediately. Foy says that since the government won't provide a record to the court, the Conservation Law Foundation has prepared excerpts for Your Honor from that record, and he offers a thick, carefully tabbed compilation as Exhibit One. Oh, says the government, "we have a massive record but didn't want to bother the court."

Mazzone wants the lawyers to keep it short. "What you have to understand is that I have a lot of work to do. The more time you give me, the better job I will do. I plan to give you a decision on Monday at nine or nine-thirty." Mazzone wants to know if the Massachusetts and Conservation Law Foundation cases have been consolidated. Doug says they have. All right.

Steve Leonard leads off. He says Lease Sale Fifty-two is the largest single-area sale in history; that it includes the great fishery of Georges Bank; and that James Watt did not do a proper job of balancing oil and fish. He says the secretary didn't give adequate explanation as to why he passed up an alternative approach to leasing, developed in the environmental impact statement, that would have exempted prime fishing areas and the lobster-rich canyon heads.

Mazzone wants to know if the commonwealth would have accepted such an alternative.

Leonard says he hasn't spoken to his clients about that, but he believes they would have.

That doesn't sound right to Mazzone. "It seems the commonwealth has conceded a great deal in this case already — perhaps too much," he says. Massachusetts negotiated with Interior and got it to delete half the tracts the state wanted deleted. Its lawyer, Leonard, is of the opinion the state would have settled out of court if a bit more of the richest fishing and lobstering grounds had been withdrawn. And yet, despite that flexibility, Mazzone says, "you are arguing that the entire sale ought to be enjoined."

Steve Leonard colors. He says he thinks the commonwealth is entitled to seek relief against the entire sale, given its other shortcomings.

All right, says Mazzone.

Leonard switches to the other shortcomings. The first one he comes to is the consistency business, the issue California employed against Watt. True, Leonard says, Watt has filed a consistency determination, under protest, in this case, but he is not meeting the requirement that the sale be consistent with the commonwealth's coastal zone plan to the maximum extent possible. "I think," Steve says, "that it is plain from the legislative history we cite that it was Congress' intention that the state programs were to be supplemental regulating requirements." Leonard is contending that appropriate parts of the Massachusetts coastal plan should become part of federal regulations governing oil operations off the commonwealth's coast.

"Sounds like you have a veto power over the plan," Mazzone says.

"Your Honor, we have it only by virtue of our having an approved coastal program that the federal government has looked at closely." Then Leonard goes after Watt's refusal to delete canyon-head tracts that seemed to have any hydrocarbon potential. That wasn't balancing, he says, that wasn't weighing the value of lobsters against the potential value of oil. Yet the law requires the secretary to strive for precisely such balance.

Leonard turns to Foy "for the remainder of plaintiffs' argument." Doug rises and says he's been losing his voice all week. He hopes His Honor will bear with him. Mazzone smiles and lifts a box of cough drops from his desk. No thanks, says Doug, he thinks he will make it. And proceeds to distance his argument as far as he can from Leonard's. Perhaps deleting some tracts would satisfy the commonwealth, he says, "but as I am sure you are aware, we are arguing that the entire sale is illegal, that it is premature, based on inadequate information that has never been seen or reviewed by the public, information that has been suppressed and never brought to the attention of the public or anyone else. I will go through that in some detail this morning."

Which he does. First, by asking Mazzone to look at page one of the tabbed exhibit. All those dry holes on Georges have forced the Minerals Management Service to reduce its estimates of what might be out there. By thirty-fold! "We have an environmental impact statement that was done in great detail, that took a long time to prepare," Foy says, leaning forward into the stroke of his argument, "for a lease sale we no longer are faced with." The defendants may argue that less oil means less potential for environmental damage. That may be so, but the puny oil patch we're talking about here may not be worth going after at all. And how can James Watt strike a careful balance between social benefit and economic damage, how can the public review his actions, "when we have an environmental impact statement for a fictitious sale?"

Just a slight pause for effect, and then Doug is asking the judge to leaf along with him through the exhibit, to read government statements about the importance of all the environmental studies it was funding on Georges. When Cecil Andrus was arguing for Lease Sale Forty-two, his lawyers insisted that "the various environmental studies which have already begun and which will continue . . . will provide a complete and tested system for analyzing and preparing for leasing operations in virgin offshore territories." And on to a chart that shows when these studies were begun and what has happened to them.

Doug wants to be sure Mazzone looks at that. "Your Honor, I have a separate copy of that so you won't have to dismantle your copy."

"I will do it," Mazzone says. "Don't worry about it."

"I do have an extra one," says Foy, waving it.

What the chart shows is that the Bureau of Land Management published the draft environmental impact statement for Lease Sale Fifty-two before most of the much vaunted studies — on whales and drilling muds and biological communities of Georges Bank — had been completed. What that means, Doug says, is that the secretary did not have the information he needed to do his balancing act. Also, the Endangered Species Act requires the secretary in his balancing to use the best scientific information available, "the best available, Your Honor." Yet when the government was preparing its impact statement for Fifty-two, it used old information in stating that there were 22,000 sperm whales in the North Atlantic and the loss of the few that might get oiled or caught in a cable on Georges wouldn't be serious. If it had waited for its own study, Doug says, it would have found that the true count was closer to 200 whales, that losses would be proportionately far more serious.

Mazzone leans back to stretch under the gold stars.

Doug has an amended complaint now, one filed two days ago. It is based on a Boston *Globe* article reporting that the Boston office of the Environmental Protection Agency had severely criticized the draft environmental impact statement of the sale but that the criticism had not shown up in the final version of the statement. "There is no mention of any EPA comments there. That is a direct violation of the National Environmental Policy Act, Your Honor, and that alone would be sufficient to justify the issuance of an injunction." The *Globe* reporter who wrote the piece is sitting in the jury box watching Foy, trying to keep his face in control.

The judge says that Foy's amended complaint is already out of date. Suits like these are filed against individuals in their capacity as the responsible officer of a particular federal agency — hence *Massachusetts v. Watt.* The Environmental Protection Agency official named as defendant in Foy's complaint has just resigned as acting director of EPA in the flurry of departures, forced and otherwise, going on at the agency as scandal mounts on scandal.

"We may be the only people to have successfully sued him before he got out the door," Doug says.

"You caught him pretty fast," the judge says, and the spectators laugh.

Doug checks his watch. The defendants will argue, he says, that this is just a lease sale, harmless in itself. "If their argument is now that nothing is going to happen and, therefore, that there cannot possibly be irreparable harm, then why should they prepare an environmental impact statement? They have accepted the proposition that this lease sale, the act of selling leases, is the significant event to which will attach irreparable harm, if the public process is ignored or seriously distorted."

Doug thanks the court, the court thanks him and suggests everyone take a little break. Doug comes over to the jury box and asks the *Globe* reporter how he likes being party to litigation. The reporter laughs.

After the break, Margaret Strand, the lead lawyer for the government, takes out after the plaintiffs in fine style. We've been through all this before, she says, through litigation and delay and finally a lease sale on Georges Bank. "Here we are again, two and a half years later, the same parties, similar claims and allegations that proceeding with a lease sale will wreak havoc on a portion of the North Atlantic coast. It is a significant resource in terms of fisheries. It is a significant resource, we hope, in terms of potential for oil and gas development. We wouldn't be here today, we wouldn't be in New York Monday attempting to conduct a lease sale, had not these very factors been considered, analyzed, and balanced."

Doug sits back in his chair, sucking on a cough drop,

watching his adversary, listening to her voice tighten as she builds her attack.

"The only issue, we submit, that is before the court today is whether or not these parties have alleged and shown any injury sufficient to warrant the extraordinary relief which they seek; and, on the law and the facts, the answer to that inquiry is a simple and flat no." Risk, says Strand, is not before the court. The government has complied with Massachusetts' request to delete tracts in less than sixty meters of water, and in most of the canyon-head tracts, and there is no imminent threat to the commonwealth since the tracts are so far from its coastline.

Mazzone asks Strand what he asked Foy: Does the lease sale constitute irreparable harm?

No, she says.

Mazzone uses Foy's point. "Then why did you go through all this procedure? Why did you draft the environmental impact statement? Why don't you wait until the drilling and exploration phase?"

Strand is quick and smart. "If we did not face litigation on every lease sale, we could look at the environmental impact in a comprehensive way at a later date." She attacks Doug's arguments, scoffing, undermining, and Doug sits and sucks. She says the scientific studies commissioned by the government are part of an ongoing process, that the agencies are constantly looking at new information as it comes in. "What plaintiffs refuse to acknowledge," she tells the judge, is that James Watt considered "the very data they now insist you enjoin the sale on." That it wasn't in the environmental impact statement is immaterial. The plaintiffs, she complains, want to "have the court create and then enforce new law that would compel the environmental impact statement to be a decision document," when the Supreme Court says it isn't.

Steve Leonard gets the treatment next. "Does the

state have a veto power?" Strand asks. "Must we accept every tract deletion of every coastal state recommended under the rubric of coastal zone management?" No, she says, and cites a recent ruling from California to strengthen the point.

Strand steps around the Environmental Protection Agency. What happens there is their business, she says. James Watt gave the agency plenty of time to comment on the impact statement.

Margaret Strand sits down, and the oil companies' lawyer gets up — Edward Bruce, Foy's friend, the one he had twitted in the Supreme Court corridor back in November of 1979 for not knowing how to type. Lease sales are not cast in concrete, Bruce says. They are processes that can last twenty years, "and we have at the beginning only the vaguest idea of what may be out there, particularly in the North Atlantic, where there is very little and, to this point, still unsuccessful exploration." Fifty-two is necessary to broaden exploration.

Rebuttal time. The plaintiffs take a few minutes to repair as much of the damage inflicted by Strand and Bruce as they can. The judge thanks everyone for keeping to the schedule. Call my office on Monday, he says, and my assistant will tell you what my order is — injunction granted or denied. He'll release a memorandum on his findings later that day. Then A. David Mazzone leaves the bench, carrying his weekend — briefs and exhibits piled a foot high — in his hands.

Granted. That's the word over the phone on Monday morning. Doug Foy and Steve Leonard and their coplaintiffs have won. Back in his bay-windowed office with his bicycle and his briefs, Doug says, "I didn't expect the judge to come down as he did on our side all along the line."

It is clear from Judge Mazzone's memorandum, released Monday afternoon, that he has read the record,
that he must have begun his research well before the
weekend. That, Doug thinks, heightens the impact of his
decision. Mazzone has found that the environmental impact statement for Fifty-two was inadequate, principally
because it didn't take into account the drastic reduction
in the government's own petroleum resource estimates
for Georges Bank; that Watt should have waited for the
completion of research like the whale study before making his move. He has found that the lease sale process itself *does* affect the coastal zone, particularly its economic
development; that Watt has failed to prove that his actions are consistent with Massachusetts' coastal zone
management program. He has found that the secretary
was arbitrary and capricious in establishing a balance between living and nonliving resources on Georges Bank by
giving undue weight to oil. He questions whether the
Environmental Protection Agency dealt properly with the
criticisms of the sale from its Boston office but says that
issue must await trial on the merits of the suit on Lease
Sale Fifty-two, in which he sees reason to believe plaintiffs will prevail.

"In short," wrote the judge, Georges Bank "represents
a renewable, self-sustaining resource for the entire nation. Because I find that the plaintiffs have demonstrated
that important safeguards directed at the decision-
making process leading up to the consummation of proposed Lease Sale Fifty-two have not been complied with;
and in light of the significance of the Georges Bank fishery resource that may be jeopardized by that sale, I find
that the plaintiffs have adequately demonstrated that
they will suffer irreparable harm if this injunction does
not issue."

"It is the most powerful outer continental shelf decision I have ever seen," Doug says. "Ever."

˙ˇ Lease Sale Fifty-two was to have been held on Tuesday, March 29, 1983, on the mezzanine of the Roosevelt Hotel in New York. The Roosevelt is bigger and flashier than the Biltmore in Providence, but the same variety of folding wooden chairs have been set out in rank and in file.

News of the cancellation has gone out earlier and faster than it did at Providence. The huge room is deserted except for some employees of the Bureau of Land Management packing up their papers, and a Greenpeace representative with a load of press kits he never had a chance to hand out. The kit has a quote from Jacques Cousteau that somehow lost its sequiturs in translation. "Each one of the cells of our body is a miniature ocean," says the captain. "Poisoning the sea will inevitably poison us." The government has left a few copies of its handouts on a table. One expresses dismay over Mazzone's ruling and confidence that "in the final analysis" — in the normal course of the appeal process — "our stringent environmental and safety controls will withstand scrutiny of the Court." Interior has decided, in view of the vigor of Mazzone's order, not to seek an emergency appeal.

A few oil people come in. "I just got the word in the *Wall Street Journal*," says one, looking over the brittle brigades of chairs. Two others are climbing up the stairs from the lobby for a look at what might have been. "How about we go down the same way we came up?" one says. They do.

The elevator doors *shush* open in the lobby, and out steps Frank Basile, well suited, neatly bearded.

"Hello, Frank," says a friend. "Seems like we've been through this before."

"About a hundred times."

When we first met, Frank told me he'd like to stay in

his job long enough to see some production. That would be a good finish to his work with the Bureau of Land Management, he had said. Now, his office is being crated up and sent to Virginia, and he is a bitter professional trying not to sound bitter. The results of Watt's accelerated program have been "absolute zero," he says. "The only difference is that a lot of people are spending a lot of money and a lot of time in court." More than likely, he says, there are going to be additional challenges to this program — some in litigation, some in Congress — until we know if the North Atlantic has any gas or oil worth worrying about. If it doesn't, "we'll go concentrate somewhere else."

And what about Frank? He isn't going to Virginia. He pulls on his cigarette. "I'm going to take my father's advice and find a real job."

8

O Canada

THIS year, the weather is the opposite of what we had
aboard *Oceanus* in the fall of '80. This year, 1983, it is
July, not January, in October. There is a gale of wind
coming, yes, but this time I am waiting it out on land.
This time, I am walking in shirt-sleeves on the roads of
Lunenburg, Nova Scotia, out along the bays, past fishing
herons, past a blue-and-white sign welcoming tourists to
"The Fishing Capital of Canada." What a hold the sea
has here, arms and fists and fingers of it thrust into the
coast. In the bright air over the headlands, there is no
sign of tropical storm Dean. He is off south'ard, bothering the Americans. With luck, he'll swing out to sea
and let the Lunenburg scallopers have one more go at
Georges Bank.

The British, before they threw the Catholic French out
of Nova Scotia in 1755, thought to limit their influence
by importing Protestants from Germany and Switzerland.
About fifteen hundred of them settled in Lunenburg.
They were farmers, and the soil of Lunenburg County
was richer than in most places along the southern coast
of Nova Scotia, where the earth is apt to be as thin over
the shield rock as meat on a shin. But the cod and mackerel and herring were so plentiful in the season of the

year that some of the farmers turned fishermen or split the difference, leaving their wives and children to till while they went to sea. The shore fisheries provided enough for export after a time, and Lunenburgers began sending lower grades of cod, salted, to the West Indies, just as the Yankees were doing.

Innovation moved slower in the Canadian fishery than in the American; shallops were still in use in the outports along the Maritimes long after the first schooners sailed out of Gloucester. In its own pace, though, and with Teutonic thoroughness, Lunenburg finally did turn shipwright. Only a couple of dozen schooners worked out of this county in the middle of the last century. By 1900, there were 150 of them, a migratory flock winging for the banks. In 1921 came *Bluenose,* the queen of the fishing schooners in the North Atlantic. She beat everything Gloucester and the others could muster against her in the international fishermen's races. Lunenburg climbed to the primacy of Canadian fishing ports. Even newspapers in Halifax, the provincial capital and, in fisheries matters, the company store, took notice. "Why, these despised Dutchmen have done more to foster the art of shipbuilding in Nova Scotia than any other class of people within our borders," wrote a commentator. "The vessels these people build are marvels of neatness. We verily believe they lavish a greater amount of money in ornamenting their craft — in carving and gilding — than they will be willing to disburse in decorating their frows [sic] and daughters."

The town still fosters boat builders. That lovely little black-hulled schooner at the head of Front Bay probably was born here, and also that Cape Island lobsterman, cuddy forward and beam broad as a shoe, heading out. He is as beamy as he is to get the most out of the law. If you're over sixty-five feet, you can't fish within twelve miles of the coast; he's an inch under and as wide as he can get and still answer the helm. It looks as if he's

headed for Cross Island. Chances are he's owner operated. Offshore is company territory, scallop and fish
draggers belonging to National Sea Products or Nickerson or Pierce. Out there, it's the highliner and the bottom
liner. Inside, it's still the man and the sea. No three-mile
limit here. In Canada, the federal government manages
right to the shore. The provinces handle the processing
and marketing ashore.

I watch the lobsterman from the height of the town,
the ridge end of the cemetery. Captains lie here in abundance, and master mariners, but the names are not German. English and Irish and Scots command the high
ground. They share the ridge with stones put in and paid
for by people still living. One of these is a marble mirror,
deeply incised with PLEASE PASS THE BUTTER. Further
along is what looks like another attempt at sepulchral sil-
liness: WHYNOT. But that turns out to be a family name.

A good number of American bones probably moulder
here. New Englanders, mostly fishermen, came to Nova
Scotia before the Revolution, and at one point accounted
for three-quarters of the population. British efforts to rein
in colonies whose trade it regarded as overly competitive
with its own fired up Nova Scotians as well as Americans.
Some towns in the province tried to remain neutral when
the fighting came, and Nova Scotian boats helped more
than one Yankee prisoner escape. But all-powerful Halifax remained a British bastion, and the Americans, ignorant as usual of the true state of affairs in Canada,
drained off the goodwill of the outports with indiscrimi-
nate privateer raids.

In the end, Nova Scotia remained loyal, and thousands
of American loyalists fled there when they saw their
cause in defeat, to found towns like Shelburne and Digby
(Digby is home port for many of the inshore scallopers).
Since then, the traffic across the border has been two-
way. Lunenburg loves to taunt Gloucester. Foreigners
manned the fleet there, they said along the bays, and few
native-born Americans found their way to the banks. The

trouble with that argument was that many native-born Canadians were finding their way to the banks in American boats. When times have been hard in the outports, which they have been often, people have gone where the shares have been higher. At one time or another, a good many men on the Boston or Gloucester docks have spoken Newfie (Newfoundland dialect) or Novy (Nova Scotian).

My waiting and wandering began two days ago when I showed up at Deep Sea Trawlers, down by Front Bay, staggering under a seabag jammed as usual with twice the gear I'd need. Deep Sea Trawlers is owned by Pierce Fisheries, one of the most energetic scallop hunters in a province that depends heavily on scallops to keep its five southwestern counties going. Pierce is a family business, started by Ernest E. thirty years ago and now run by his son, Forrest.

The manager of Deep Sea Trawlers, Gerald Penney, a stocky man with a crew-cut beard, looked up from his desk and shook my hand. "Well," he said, "I see you finally made it up."

"After fourteen months," I said. It had taken that long to find a date that weather or some other agent of fate didn't scuttle.

Gerry was talking to one of his captains. I waited in the big room outside his office, a chandlery for fishermen, especially those from Pierce's boats berthed at Lunenburg. They looked fairly young to my eyes, none much over thirty. They carried themselves as men do who know their strength and take pride in that. They wore midcalf leather boots partially or fully unlaced, as if they were so used to seaboots they didn't want to mess with thongs. They jollied the man behind the counter and ordered cases of soft drinks for the twelve-day stint on Georges. They bought T-shirts, sex magazines, extra pairs of tough green plastic gloves to keep their fingers from the destruction of picking and shucking scallops.

Gerry came out to take me down to the dock. I'd be going out on the *Ernest E. Pierce.* "She's new and steel, a hundred and ten feet," Gerry said. "Think you'll like her better than the wooden boats." We dodged around a forklift that was carrying a huge scallop rake to one of the boats. "They're rollers, those old ones," Gerry said. "And they're kinda ripe."

The *Ernest E.* was built in 1981, but she looks ten years older in the waist, where the rakes have hammered the hull. Inside, she is trim and remarkably clean. The smell of ocean flesh floats in the air, but without the trace of ammonia you get often aboard the fish draggers. The bridge is well windowed, and you don't have to turn your head far to see most of the navigational or depth-finding equipment you need to see. The boat cost about $2,000,000 new, but it would take about $3,000,000 Canadian — worth $2,250,000 U.S. at this writing — to replace her.

Gerry and I stood in the wheelhouse waiting for the skipper. This could be the last trip until next March or April, he said. A little early to quit, but expenses were just too high and scallops too scarce. The boats had been going "out east'ard" to give Georges a rest, steaming as long as forty hours to reach pockets of scallops in Middle Bank, Banquero, and on along the chain of rises leading to the Grand Banks. Lunenburgers use the lay system. The owners take 40 percent of the proceeds from a trip, the fishermen 60 percent. Pierce pays for 40 percent of the fuel, all repairs, all electrical and electronic equipment. The crew handles the rest of the fuel and the provisions. The *Ernest E. Pierce* needs to come home with about 14,000 pounds of scallop meats to cover expenses and the bangs and breakdowns the sea usually hands out. Maybe 12,000 pounds would do, if the price holds at the astronomical level of $5.75 a pound dockside. But last trip, the boat came in with 10,000 pounds. "We can't afford this," Gerry said. "The men working the boats can't afford it."

Georges is the prime scallop ground, and scallops are the prime sea harvest for Nova Scotia, which regards itself as first among ocean fishers in the Maritimes, or, for that matter, in the whole western North Atlantic, from Cape Cod northeast. Scallops made up a third of the total value of the province's landings in 1979, as against 18 percent for cod and 16 percent for lobsters. But Georges is disputed territory and has been since before 1977, when Canada and the United States pushed their fisheries jurisdiction to 200 miles. Washington claims all of the Bank, Ottawa the eastern third. While the case is before the World Court, the two countries have agreed that both may fish in the contested zone.

Pierce scallopers used to go as far south as Virginia once in a while. "Now," Gerry said, "we're like a farmer who has had his farm cut in half and yet he has to grow the same amount of potatoes. You just can't do it, and that's what we're facing on Georges Bank."

Gerry spotted the skipper of the *Ernest E.* on the dock, unlaced boots and all, gamming with the other captains going out with him. "Harold Moore is good," Gerry said. "Few years back, a man like him could pick up over a hundred thousand dollars in a year of fishing. He'll probably make sixty thousand this year." Deckhands were making $50,000 in the good times, he said, and now they're down to half that.

Moore clapped a friend on the shoulder and vaulted up to the bridge wing from the dock. He is short and broad and wears a helm of shining brown hair and a smile. His face is Irish. He began to speak, and at first I feared my deafness had worsened in the alien weather. Not at all. Moore and most of his crew speak Novy. It sounds a bit like Maine gone the full distance eastward, Bangor in Dublin. I said something to Harold and jumped a little at his response: "Eye-eeeee," a long and rising sound. What he was saying was: "What?" On the third try, he got my message. I was asking what he called the steel-framed wooden contraptions hinged to the gunwales in the waist. "Dump tables," he said, with his smile. We had connected.

Word was out that there was some kind of writer aboard *Ernest E.* Doug Dory, Moore's mate, peered at me for a while. "Gettin' it all down?" he asked.

A big crewman with a moustache hanging like baleen over his mouth peered at my notebook scribbles. "Writin' a story, are you? We'll give you a story. Big storm comin'."

"I thought the one that was out there was heading out to sea."

"Couple warm days, it'll haul in on us."

The skipper of the *Angela Michelle,* moored across the dock, shouted at me. "If you run out of answers, come aboard and we'll give you the real ones." He laughed and looked at Harold Moore and shook his head. Then he stopped smiling and said, "I dunno, I guess it's the same hell either way."

The town rose above *Ernest E.*'s high bow, first the barn-red sheds of the boat builders and ship fitters, then the pastels and whites of the homes, many of them Victorian. On the top of the ridge, next to the cemetery, stood a castle of gingerbread, the county academy building. "Monday," Harold was saying to a young crewman with a Gaelic face and one of those super-decibel tape decks under his arm. "Monday it is." The captains had decided, after almost two hours of hearing each other out. Not many wanted to steam fifteen hours to Georges and take a pounding from tropical storm Dean. Today was Friday. Boats don't go out on the weekend, so it was Monday. Crewmen picked up their bags and hustled over the gunwale toward two gift days ashore. I hoisted my leaden duffel and climbed the hill, cursing my luck. I needn't

have. I found an inn with fine food and a Victorian view
of the bay, of the *Ernest E.*

Canadians and Americans have been sharing and
squabbling over the same fisheries since long before they
were Canadians and Americans. The American colonists
began building a fleet for the Grand Banks run that
would eventually rival Britain's. They complained regularly about harassment from their competitors and once
sent a contingent of militia into Nova Scotia after the
French. They wrote their mothers from Barbados, saying,
"Fish at present bears a good rate by reason of the Newfoundland men are not come in." After the Revolution,
the American fleet decimated, they wrote Ben Franklin
and his colleagues in the peace negotiations with Britain,
insisting, by virtue of their former Britishness, on their
continued right to the Grand Banks fishery. They wrote
to George Washington, saying, "The New England states
particularly cannot do without it, for it is their whole and
sole staple worth speaking of." The American negotiators, listening, came away with the Treaty of Paris in
their pocket, with Britain's acknowledgment that New
Englanders could not only have their whole and sole staple but cure it "in any of the unsettled bays, harbors and
creeks of Nova Scotia, the Magdalen Islands, and Labrador." British fishermen received generally the same
rights in American waters, but that just didn't get the
same play in New England.

The pattern has held, off and on, ever since. No real
fighting, you understand. There has been none of that to
speak of ever since the Americans went stumbling
around in the northern wilderness during the War of
1812 and then thought better of it and came home. The
differences are troublesome, nonetheless. That is how

the Canadians regard the dispute over Georges Bank. In the late 1970s, their Secretary of State for External Affairs called it "the most serious issue we have with any country." Statements like that are Ottawa's way of trying to get Washington to listen for more than the customary minute. Canadians often feel in that regard like the farmer who was obliged to belt his mule between the eyes with a billet of firewood — just to get its attention. (The metaphor is apt, but the choice of animals may be off: Pierre Elliott Trudeau, Prime Minister of Canada during the thick of the fight for Georges, said that living next to the United States is like sleeping with an elephant.)

There is more to the Georges Bank quarrel, too, than just David vs. Goliath. The argument went to the International Court of Justice late in 1981, and its resolution will affect the scores of maritime boundary disputes resulting from the new worldwide urge to declare 200-mile ocean jurisdictions. What happens on Georges will tinge what will happen when the neighbors on either side of the world's longest undefended border try to settle other even more important problems, from the allocation of water resources to acid rain. After all, as the Canadians like to say, "The Americans are our best friends, whether we like it or not."

The Canadian Department of External Affairs is sited in a curve of the Ottawa River, as light and pleasing a structure as the Department of State building, gloomily looming in Foggy Bottom, is not. Officials there are much like their counterparts in Washington, though, in saying as little as possible about their legal strategies for Georges while the case is at bar. They profess anger. "The United States wants all of Georges Bank," says one senior man. "There is no more rationale than that."

The move that triggered the standoff was Canadian, as they admit. In 1964, they began issuing permits to Canadian oil companies and Canadian affiliates of U.S. oil companies to explore the thumbnail of Georges Bank, the part Ottawa claims is Canadian. The permits were issued up to a line drawn equidistant from points on the Nova Scotian coast and points on such places as Cape Cod and its neighboring islands, based on rules agreed to by an international convention in 1958. The Americans, when apprised, did indicate that they might question the precise lay of the line. But, says a Canadian official, "it appeared that they would accept an equidistant delimitation." (American authorities now strongly deny they sent such a signal.) In 1968, the Canadian official continues, the State Department asked Canada to impose a moratorium on oil exploration in the area, but that seemed to Ottawa to have more to do with environmental than jurisdictional considerations. Not until 1969 did Washington formally inform Ottawa that it did not accept the validity of the Canadian permits and ask for negotiations on the matter.

The two governments were as yet dealing only with the seabed and what lay under. On the West Coast, both the U.S. and Canada had issued oil-hunting licenses up to a boundary equidistant between British Columbia and the state of Washington. But as the boundary negotiators met, the United States intimated not only that it rejected equidistancing in the Gulf of Maine and Georges Bank, but that it was thinking of laying claim to the whole Bank. When the 200-mile limit went into effect north and south, says our source in Ottawa, "the Canadian government first saw the American line on Georges. It didn't come as too much of a surprise to us that the U.S. claimed the same area for its fisheries zone as it had for the seabed."

From feuding over a resource that might or might not lie in the rocks of the Georges Basin, the United States and Canada turned to a fight over an actual resource whose importance, especially in its northern range, went far beyond the economic. The Canadians talk in terms of need. They say they need fish to ensure the survival

of their Atlantic coast communities. Unlike New Englanders, they say, their outporters can't simply move to
jobs inland if their catches fall off; there is no safety net
inland. The Carter administration listened to them and
agreed to negotiations on the entire boundary-fisheries
problem, to be carried out by the best men each could
find.

Monday we go. Down at Deep Sea Trawlers, the man
at the counter handles a last-minute buying spurt: pop,
copies of *Hustler,* one or two newspapers. Six boats are
going to Georges this time. The catch there is about the
same as it is out east'ard, poor, but less fuel will be spent
in getting it. Someone says he's heard the price of scallops may be up, to six dollars the pound, by the time *Ernest E.* steams up Lunenburg harbor twelve days from
now.

Cars drive down to the dock. Most wives and kids stay
inside, talking to the men through open windows. One
deckhand brings his young son aboard *Ernest E.* in his
arms, and later the boy, alone, swaggers a small swagger
in his tiny seaboots back to his mother. Doug Dory, the
mate, sits by the starboard window in the wheelhouse,
squinching his broad mouth from time to time to spit
over the side. A three-masted square-rigger has come in
over the weekend. Doug says he has heard it's an Austra-
lian training ship with women cadets aboard. "The other
half's kangaroos," he says, and chuckles, and spits.

It never fails. When a vessel lies tied up but ready for
sea, she is a thing apart. A dozen feet separate the crew of
the *Ernest E.* from their dockside friends, their fellow
scallopers. But by being on board, they have, in an eerie
way, already taken their leave. Then the lines go. "All
clear," Harold Moore sings to himself, and takes his boat
into the harbor.

We pass National Sea Products' plant to port. Four or

five big stern trawlers lie along the docks. Louis Lantz, the cook of the *Ernest E.*, tells me we're looking at the largest processing plant on the western North Atlantic. Like other large fishing firms in the Maritimes, National Sea is in financial trouble. The federal government has just announced plans to follow the recommendations of a special commission and both invest in and restructure the companies. This has caused much noise in Nova Scotia, as conservative as any New England state. My copy of the *Halifax Herald Limited* has an editorial roaring at a federal establishment "that would like to put a stranglehold on the same private sector it induced by monetary and regulatory means to acquire a catching and handling regimen that cannot bear either the weight of time or the potential of the sea." Ottawa, said the editorial, did not seem to have "bothered considering the relation of the U.S., that huge market to the south, where the private sector loathes the prospects of competing with any foreign government-subsidized enterprise, and where the reaction most often is a countervailing duty."

At the Pierce dock fifteen minutes ago, Gerry Penney had delivered his own gripe. Those big companies usually set scallop prices, he said. "If they're going to be government sponsored, they're not going to pay much attention to those prices. If they lose on scallops, they'll get a government grant to make up for it. It's hard to compete with that."

We pass a touch of October among the trees on the starboard point and head out into a school of islands. A swell from some storm, Dean perhaps, lazes in and crashes on the headlands. *Ernest E.* shakes her bows, and I know then that I am riding a fair roller. I go easy on the cold cuts Louie Lantz has set out for lunch. The crewman next to me says he's forty, but the loss of his lower teeth makes him look older. He says he's been fishing for thirty years and he's never seen scalloping so bad. Maybe it was, twelve or so years ago, he forgets. He hears

the scallops are at the bottom of a cycle of six or seven
years or so, but he doesn't know about that. What he
knows is the scallops are so small, "they grab you when
you go to shuck 'em."

I spend the afternoon wedged into my corner on the
bridge, looking out at the most common commons of
them all and listening to the skippers and mates on the
radio. I am used to the chatter, but this is almost confes-
sional. Men get on the horn and stay on, gamming, gos-
siping, taking sonic solace. "Fits and starts, skipper, fits
and starts," says one. "My, my, my, Jesus Christ!"

Around one o'clock Harold Moore heads for his cabin
aft of the bridge. "See you in the mornin'," he says. He
will say the same thing about the same time every day,
and we will laugh with him. He'll get four or five hours'
rest, then relieve Doug for the alternating six-hour
watch, a standard schedule when the scallops are running poor to fair. When they run in abundance, it's eight
hours on and four off for the deckhands. No chance of
that this trip, though we hear someone on the radio
claiming he has iced off two hundred bags in four days.
That's pretty good, if it's true. He isn't saying where
he is.

About 180 miles off our port quarter as we run down
the coast is Sable Island, home of ponies and shipwrecks
and a few people and some exploration rigs. Its waters
support a good fishery, but gas is what makes the news
from there now — the promise of gas from Mobil Canada's Venture find. Distribution companies are already
talking about piping Venture gas to New England; there
isn't enough demand in the Maritimes to create much of
a market for it. But first, Mobil and other companies must
find billions in the banking system to ready the field for
production.

Dead ahead now is the tip of Georges. Doug Dory looks
at my map showing the drilling permits that lie under
moratorium while the boundary dispute wends its wordy

way at the Hague. The blocks are much larger than those offered off American coasts. When Canada began leasing in the Atlantic twenty years ago, it was extremely generous with its territory and with its permit regulations. It had to be to compete with the North Sea, then the oil gambler's hottest game. Competitive bidding didn't apply. The Canadians, like the Europeans, sat down with applicant companies they thought could do the job best and gave them ample room to do it in. In recent years, the trend has been toward renegotiation, smaller tracts, and a higher proportion of Canadian ownership in ventures moving from exploration to development and production. But even now, tracts of several hundred thousand acres are not uncommon.

Doug runs his finger across holdings on hold, colored and crosshatched geometries covering most of Georges' thumb tip. The names in the legend are Dome, a vast Canadian company, and the local affiliates of American giants like Texaco, Chevron, Mobil, others. "As far as I'm concerned," Doug says, "if they find oil there, that'll be the end of it."

My bunk, as usual, is the upper one. It hangs high in one of the crew cabins aft, just forward of the big diesel. Dismantled cardboard boxes cover the floor to separate drying seaboots and the tile. On the bulkhead is a poster from one of the skin magazines — breasts, vulva, and all. And finger marks. Scratches on the dungeon wall. Some men spend half a year out here — twelve days out, four or five in port, twelve out, from March through November or early December. The only sustained time they have for real women is during the winter, when they can draw unemployment insurance and relax. A captain in Woods Hole called scallopers animals, and I am mindful of that as I wait to make the acquaintance of my cabin mate. He turns out to be young and human, with a smashed nose and a mild outlook, at least toward me. "Old boy," he calls me. "Supper, old boy?" he says, shaking me out of my nap. I go on deck to let my senses rise from sleep and look eastward at the fang of a moon in a greening sky.

Just before midnight, the diesel dwindles from gnashing to murmur. We're on the grounds. I climb to the wheelhouse to see the mate's thick body silhouetted against the blaze of the decklights. The winch cables are coming in, up from the water, through the huge blocks in the gallows, around great steel idlers forward and then aft to the winches. Deckhands fasten hooks from the booms to the dump tables, and the great slabs of wood and metal tip up and overside. Rocks and trash from the first haul splash into the sea, and the tables are set gently down on their wooden deckpads.

The rakes break water. Boom hooks are fitted to their noses, and they ride up into the night, ten feet or so of chain, steel links, and thick nylon mesh. The rakes — dredges in New England — are around fifteen feet across, and the rectangular opening into the bag is a foot or so high. More than two tons of metal hang from each boom, and the thick masts quiver with the jerks of the winches. The rakes crash onto the dump table, and only then do the deckhands move into the waist. There is no room to duck away from those monsters. The men guide the noses inboard and down and reset the boom hooks to lift and empty the bags.

At first, all I can see are rocks, from pebbles to boulders, and a few monkfish and skate. But the pile has a tan look to it, and in a few seconds I make out why. The rocks are nested in scallop shells, a clutter of trash from years of fishing these bottoms and from natural mortality. Here and there are live ones. The men start picking, four to the port dump table, three to the starboard. They are one short this trip, but not for long. Doug Dory grabs his gloves and hurtles down the ladder to the waist, delighted to get his head between his knees and sort through the ruck. He has spent nineteen of his thirty-three years in and out of this position, and through the trip he'll make the dash from the wheelhouse down to his beginnings whenever he thinks he can.

The men fill wire bushel baskets identical to the ones on *Delaware* and *Valkyrie* with scallops ranging from perhaps three to perhaps eight inches across. The sound is crisp and strange as they work their way in stooping frenzy through the piles; when I close my eyes I hear the crackling of ice. In ten minutes they are done, only a couple of bushels picked from each dump table. The rakes are working again. Once emptied, they were swung outboard. When both were in position, Doug tooted the horn, two deckhands swung mauls down on the releases, and the rakes crashed back into the sea. The scope is the same as in trawling: you let out wire until its length is three times the water depth you're fishing in, plus ten fathoms for good measure.

We can see the lights of other scallopers. Doug peels off his dripping gloves and calls one. "That you, Douggie boy?" says his friend. Doug says aye, it is. "There's no big ones, and the little ones are gone," says the voice, blasting full volume through the speaker. "Wasted three days out east'ard and come here." A long silence. "Jesus, Jesus, Jesus, I'm so fed up with this. There's no future in it."

Running fore and aft below us are corridors between the galley and the hull. The men working each dump table have carried their baskets here and put on aprons and grabbed short, curved shucking knives. Each has a clean bucket for his meats. They begin a standing dance to keep their shucking at speed, rocking back and forth twice for each scallop. In the first rock, they open him and throw his innards and half his armor against the hull wall. The refuse drops into a sluice of seawater that carries it to a drain and gushes out and down to the waiting birds. In the second, they cut the muscle from the remaining shell, flick the meat into a bucket and the shell into the sluice. One, two, in the space of four or five sec-

onds. A couple of men wear the white cotton sacks used
for storing the scallops, forty pounds to a sack, in the ice.
They have them rolled up and jammed on their heads as
caps. The smell is of salt, and of sperm.

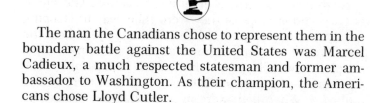

The man the Canadians chose to represent them in the
boundary battle against the United States was Marcel
Cadieux, a much respected statesman and former am-
bassador to Washington. As their champion, the Americans chose Lloyd Cutler.

Cutler is an international lawyer so busy that he keeps
a half-dozen briefcases by his desk in Washington. He
moves with the mighty. He consults with presidents and
spends his vacation time in what was the summer White
House of Grover Cleveland, in the resort town of Marion,
Massachusetts, just across Buzzard's Bay from Woods
Hole. He is a solid, midsize man who projects a remarkable combination of intellectual energy and emotional
detachment.

By the time Cadieux and Cutler and their teams sat
down to reason together, their governments had come to
some interim agreements. Traditional patterns of fishing
would continue. Each country would be responsible for
enforcing its fisheries laws on its nationals within the
disputed areas. Both would see to it that no one else
fished there.

The special negotiators were supposed to work on all
the maritime boundaries — Georges; British Columbia–
Washington; British Columbia–Alaska; Alaska–Yukon
Territory — and to reach agreements that would manage
commercial fish stocks sensibly across those lines. They
spent nearly two years at it. When they found themselves
closer to agreement in the East than in the West, they
concentrated on the North Atlantic. They traded American squid off New Jersey against Canadian redfish off the

Maritimes. The Canadians got three-quarters of the scallops on Georges and the Yankees 80 percent of the groundfish. In 1979, the negotiators submitted two linked instruments to their governments. One set up a joint commission to oversee fisheries management, including the setting of yearly catch limits, in the disputed zone, and binding arbitration if things got out of hand. The other was a treaty sending the boundary quarrel to the World Court at the Hague.

"Had we been able to obtain ratification on distribution of shares of the fish catch," said Lloyd Cutler, sitting quietly in the sun beside his house in Marion, "the whole issue of where the boundary ought to be would have diminished in importance." They weren't able to. The United States Senate let the matter sit for a year before holding hearings and for another year bottled it up here and there. Finally, a group of senators asked President Reagan to separate the twins. The Canadians were outraged, or said they were. "It is particularly galling," said one of their diplomats, "to have to negotiate such treaties twice — once with the appointed representatives and again in Congress." But they agreed in the end. The fisheries agreement died and the boundary issue went to the Hague.

Cutler thinks the Canadians overtraded their position, that they could have made more concessions. But, says our source in Ottawa, "the agreement was by no means satisfying to our people. I think that on the whole Canadians are more conservative than Americans, more security minded. For us, a loss would have been much more devastating than for you. We were left in a position where we had to decide whether to go for a boundary adjudication without the security of a fishing agreement or whether we would wait and see." Ottawa, which keeps tight controls on the number of Canadian boats that can fish offshore, came to fear the open-entry attitude in the American fishery and the growth of the American fleet more than it feared the possible defeat of its claim to eastern Georges. It went for adjudication.

Regional opposition gutted Cutler's efforts at negotiation for the nation as a whole. Some New England fishermen supported him — skippers in Boston and along the Maine coast, mostly. But New Jersey fishermen didn't want to give up any of the squid they might someday find profitable. And the New Bedford scallopers and draggers didn't like any part of anything. The manager of one New Bedford boat-owners' association says, "We met Lloyd Cutler in Marion, and it looked as if the government was creating a tremendous giveaway of all our rights to the fishery. We told Cutler we were going to use whatever method we could to fight it." The method was to set up the American Fisheries Defense Committee and raise enough money from around the country to hire a high-priced lobbyist.

As Cutler himself sees it, "The fishing case was of importance to New England senators and representatives because we only had a New England deal. The rest of the Congress simply followed the lead of the New England members. It's the theory of bread cast on the waters; the other members will need the support of the New Englanders on some other issue in which the New Englanders aren't interested. And the New Englanders respond to so small a constituency as the fishermen because it's all plus; it's not like abortion, where you have one strong faction against the others. Even a fellow like Claiborne Pell of Rhode Island, the senior Democrat on the Senate Foreign Relations Committee and on the whole a very reasonable man on international issues, was plain in saying [to us], 'Unless you win the support of my scallop fishermen, I can't support this.' "

As for the Canadian and American fishermen, Cutler said, they remind him a bit of the Catholics and Protestants in Northern Ireland. "They're just too interested in their historic grievances against one another."

The memorials, the initial briefs, were submitted to a five-man chamber of the World Court, followed by countermemorials. Oral arguments took place in the spring of 1984, and a decision was scheduled for August — probably to be followed by a string of supplementary decisions on technical issues. The maneuvering has been all it should be. When a maritime boundary decision in a European case gave half-weight to a long and thin promontory that might otherwise have unduly skewed an equidistance line, Canada came out with a revised line that did the same to Cape Cod and thus increased the acreage of its claim to Georges. When another decision gave weight to the general trend of a coastline, the Americans rushed to publish a boundary that would have zipped right through Nova Scotia if the considerate claimsters hadn't put a jog in it. The Canadians, in their arguments at the Hague, stressed the economic and social importance of the Bank to Nova Scotians. The Americans argued that Georges was a geologic continuation of their mainland, having once been a part of what is now New England; that they had fished it longer and had done most of the survey work on it; and that it made ecological sense to preserve it as a unit — an American unit.

Most of those I asked thought that the court would draw the boundary somewhere between the original Canadian and the original American line; that kind of compromise has been to its liking in the past. "Wherever the border is set," says Lloyd Cutler, looking out across his summer-sere lawn at Marion harbor, choked with ocean racers readying for some do, "the fish won't know where it is. And if some part of the Bank ends up in Canadian hands, then all the issues we worked on remain to be worked out in the future. Meanwhile, the possibilities of conservation, of rational management, are all out the window."

The next morning, skipper Harold Moore is on the bridge. The air is warm and muggy, and the men are working in T-shirts. The radio talk never changes. Someone is doing a little better than usual. He shot away, Novy for turned or followed, on the 2150 loran line, and just got three and a half bushels on each dump table. I ask Harold to show me the Canadian line across Georges. He rummages around among his charts and can't find it. "Don't think I've ever seen one," he says. "They just tell us not to go further down than the twenty-six-hundred line."

Not too many birds keep us company, and most seem to be of one species. They have a dark cap on their heads and white cheeks and a white slash across the upper tail, and they spend a lot of time running across the water after bits of scallop guts. Doug calls them hagdons. The bird books prefer "greater shearwater."

One of the crewmen moves in to help the winch man turn the starboard rake so it will come over the rail with its bottom facing inboard. A chain snaps loose and cracks him across the upper arm. He crouches, openmouthed in pain, and then staggers into the fo'c'sle. Doug has just come up to relieve Harold. "It's Jimmy," Doug roars, heading for the bridge door. "Shit," he hollers. But here comes Jimmy back to his station, rubbing his arm, grinning a little. Doug closes the door.

Harold has been working around Northeast Peak this morning, just about where *Delaware* tore up her net. Now he wants Doug to try deeper water, out toward the channel, in seventy or more fathoms of water. The rakes come in, the giant reticules spill trash and tiny scallops. Doug spits sadly out the window. He looks down on the butts and backs and the hands picking. The radio says something about fast-running tides in Northeast Channel. "Used to be," Doug says, "when the tide was goin' you'd get more scallops and less crap. But now it don't make no difference." Spit. "My, oh my, oh my . . ." trailing off out the window. Then he perks up a little. Some of those scallops, he says, are seed that weren't here a couple of months ago. You're not supposed to take scallops under three and a half inches, but nobody is too concerned with that out here. They'll be averaged out at the Pierce processing plant to thirty-five meats a pound, the maximum count for export to the United States, and then loaded onto Pierce trucks for the haul south to Boston or the new markets Pierce is opening in Baltimore and down in Georgia.

Doug watches as the boom hook falls out of the eye on the nose of the port rake and a roll brings the rake slamming down on the deck. "That," he says, "is how a lot of fishermen get killed."

Doug hears a familiar voice on the radio, the skipper of a New Bedford scalloper he's seen working these waters over the years. Don't see too many Yankees around here now, he says. They're fishing further west, closer to home. Doug has been down to Maine, but the place he loves is Disney World in Florida. He spent one vacation there with his wife and young children, and he wants to go again.

That puts me in mind of what Gerry Penney was telling me at the dock. He'd been down to New Bedford a couple of times, he said, and he didn't get any hard words from the younger fishermen because of where he came from, though some of the mossbacks told him they didn't have time to talk to any Canadians. "Our fishermen know yours, and vice versa," Gerry said. "We need something, we're not afraid to ask, and you do the same to us. I think if left to themselves the people in the fishing industry would have got together on both sides, and the thing would have been settled by now. Today, they're throwing political snowballs back and forth." I asked Gerry what would happen if the United States got what it wanted on

Georges. "It'd close us up," he said. He squinted. "It would be one helluva battle."

The meals are quick and tender. Louie the cook is good at his trade, frying moist steaks and chops. No fish, no scallops. Lots of ice cream. After the noon meal, dinner, he skips up to the wheelhouse and jangles a tin can with some pennies in it in front of a radio mike. I can imagine what that does to listening ears. "Alms for the poor," Louie croons.

"You goddam fool," says a voice and breaks up in laughter. "How're yer scallops, Louie."

"Last ones were so small I thought they were prayer beads."

We settle down onto a track in the sea, following four scallopers around in an ellipse. Rocks, scallops, bitching. *Forrest Glen* steams by and asks Louie what he's got for dinner tomorrow. Louie lines out a disgusting menu. "I'm coming over," says *Forrest Glen*.

"I spent two years on that boat," Louie says to me. "God, could she roll. Took two hours to go one mile in a gale."

We are dragging at eighty fathoms and doing a little better. The scallops are bigger, but so are the rocks. Harold has the watch, and he is explaining about licenses. Each fisherman pays $20 a year for a green card encased in plastic, and the government inspectors will fine you $300 if they catch you without one. Boat owners pay $1500 for approximately the same thing. Nobody knows what those boat licenses would bring on the open market. Probably they're down a bit now that scalloping is off, but Harold thinks they'd still go pretty dear. They were out of sight in 1978, when you could make your thirty-thousand-pound trip limit in four or five days. You could ice off more than a hundred bags a day then. "Twenty now," Harold says. "You still get a few good hauls in the spring, when the scallops have had a chance to come back, but . . ." He puts his head down to the hood of the radar and lets the words hang. It looks to me as if we're going to ride right up the stern of *Forrest Glen,* run our bow right through the red maple painted across the bulge of steel. Harold nonchalantly takes over from the automatic pilot and shoots away, whistling soundlessly.

"Say, Harold."

"Eye-eeeee?"

"Where're you?"

"Inside here."

"Don't know what you're doin'. I'm not doin' fuck-all."

Doug Dory takes over. He absorbs the blasts from the radio for a while. "Bill, old boy," he says. "There's seventy-four scallopers in this fleet. We keep it up, there'll be enough scallops for two trips apiece per year."

"I see, I see, Cap," hollers someone to someone else in the formation. "I see, Cappy boy. Yes, yes, yes, yes . . ."

"We landed eighty million dollars in nineteen eighty-one. That's big industry here. But nobody pays any attention to scallops." His voice rises. "They never have. They're always talkin' groundfish, but we're right on the edge."

"Jesus Cockadoodle Christ," shouts the radio.

"Get those bureaucrats out here in a blow for twelve days, and they'll know scallopin'," Doug says. "Those Americans . . ." He's really on a tear now, but he checks himself, looking at me. "I know you're from the States," he says, "but they come up here in good weather, and when it's bad they go down to the beds further west, and we can't. We gotta stay here. And when we're through with the fuckin' around over the boundary, we'll be stuck with this."

"You can't get 'em, you can't get 'em, that's all I know," says the radio.

"Hell," Doug says, "if we can't go below the twenty-six-hundred line, the Americans shouldn't come up here." He starts looking for his pipe, finds it, sucks on it,

frowns, takes it apart. Louie the jokester has stuck a
match up the stem. Doug growls, pulls the match out.
"There's lots of scallops off Labrador, but Labrador is
part of Newfoundland, and Newfoundland is like another
country. They won't let us go there." He has his pipe
going. "They'll have to close this place for a couple of
seasons, to the Yankees and to us. If they don't, the Bank
will be dead. If all these small scallops are fished out next
year, that will be it."

Doug has worked aboard trawlers and remembers
what happened when the boats filled their limits. "They
were hauling back and throwing haddock overboard.
There were dead fish all over the place." That's what the
Kirby report says, I tell him, and I haul out the two thick
volumes of a Canadian government commission's report
on the Atlantic fisheries, published at the end of 1982.
The industry is in deep trouble, I read to him, and it is
"part of our culture as residents of a country with one of
the world's longest coastlines, a country that fronts on
three oceans." The report quotes Garrett Hardin and his
fable of the commons and criticizes the Canadians for
expanding the fishing industry in hopes of a bonanza
when the 200-mile limit came in. The result was the
same off the Maritimes as it was off New England: too
much effort, too few fish, rising costs, falling incomes.
Some stocks are coming back now, the report says, and
Canada is still the world's largest exporter of fish — half
of it to the United States — but the industry is "mired in
financial crises, plagued by internal bickering, beset with
uncertainty about the future and divided about how to
solve its problems."

The report recommends restructuring of some companies and a shift in regulations that will make it possible
for more fishermen to earn a decent living. Social management runs throughout — even a few ideas about
vesting rights to fish in fishermen. But in the end, the

report makes a sad admission: most of what it is recommending was recommended by another government commission fifty years ago. In fact, over a hundred commissions have studied Canadian fisheries in the past century.

Doug says that being the case, he's not too sorry about not reading the report. "The skin magazines are about it for me." I tell him some of what the report says about Canadians being willing to sell to the middle of the market rather than developing high-quality fish products, as the Icelanders and Norwegians have done. "You talk about quality control," he says. "I worked in a plant where the pollock were so ripe you could pull the backbone out of the meat. Hell, Canadians got used to sending yellow stuff with no taste to it to the warmer countries. After a while, they quit buying. There was better fish around, and they bought that."

I tell Doug that the Kirby report says scallops are doing well enough so that there is no need to spend time reporting on them. "Jesus Christ," he says. "What did I tell you? Jesus . . ." Anger drifts out the window with the pipe smoke as the mate glares down into the sea.

In my berth at night, next to the diesel, I dream I am surrounded by regiments of Zulus beating their shields. In the morning, Harold tells me we tangled with *Angela Michelle* last night. The rakes hooked as she came across our stern, and it was a couple of hours of backing off and turning and hauling — and splicing a lot of cable. It's the same hell, *Angela*'s skipper had said.

The long delay back in Lunenburg has put me on the outside edge of my schedule. Harold has been calling around to see if somebody's going back and hears Captain Frank Strawbridge of the *Jean N.* will be heading in to Riverport, ten miles from Lunenburg, tomorrow afternoon. Strawbridge calls after supper to say she's a few miles off and would I like to transfer now or wait until tomorrow. Doug and I agree it's better to go for it now so we can have a second chance tomorrow if we can't make this transfer.

The swell coming in from the sun is gentle. Harold takes the watch. Doug and a crewman put my gear in the ship's boat, a sharp-sterned, lap-straked Lunenburg dory. The dory tips in the air, and I almost lose everything. It rights, hits the water, and drags forward under the pipe leading from the starboard shucking sluice. A gout of scallop guts and seawater soaks my duffel.

The swell is a lot bigger when we're in the dory and pulling away. Doug says he doesn't swim. Harold waves and smiles. We pull in under the blue wooden bows of *Jean N.* The crewmen above us are riding Doug as he tries to bring us around. The stem of the dory hits the scalloper a whack and it sounds as if the stem has given. I scramble up into a scrimmage of grabbing hands. My gear flies up from the dory, and two crewmen lug it up to the wheelhouse for me. Frank Strawbridge, the skipper, is sitting there, sharp-faced and unshaven. The mate, John Colson, big and bull-necked, smiles quietly at me.

"You sleep there," says Frank. He points to what I take to be his bunk, in a small cubby about where Harold's comparatively palatial quarters are located in *Ernest E.*

"But it's yours," I say.

Frank pats my shoulder. "There," he says. There, I go.

"You came out on the newest, you're going back on the oldest," I remember Doug saying. The boat carries more varieties of smell than *Ernest E.* She has no showers, and the results stop the breath first climb down the ladder. She is perhaps ten feet shorter than *Ernest E.* and she rolls more. Her gear looks to be in good shape; the new electronic plotter in the wheelhouse enables Frank to come back to the good grounds without much trouble. But she is twenty-one years old, a converted side trawler, with only half the power of *Ernest E.* I step out on the stub of a bridge wing and wave at my old friends across the water. They wave back. Louie comes on the radio. "Ice cream for dinner tomorrow," he yells.

The wheelhouse is hot, the cabin hotter. The engine stack passes up just a few feet away, but the clatter, compared with the big Caterpillar I've been bunking with the past few days, is subdued. The sun is a bubble of blood on the rim of the sea. It is ideal weather for a green flash, that rare pulse of light the second after sunset. I do not see one now, just a scalloper moving into the sun's space.

There are sixteen men in the crew, three fewer than on the *Ernest E.* They are all young, and they all have channel fever; they hear the call of the home port. They throw rocks at each other, at a crate bobbing by under the work lights. A youngster welds a broken link in the back of one rake. He uses no protective goggles. As is her custom during the end of a trip, *Jean N.* fishes one rake at a time. Frank Strawbridge says he is working on seven thousand pounds for the trip. He turns in, and big John and I watch the stars bounce with the roll of the ship. John says he started scalloping in 1964, then did some work ashore. "I came back into this racket a while ago," he says. "Don't ask me why, I don't know." He looks at me for a while. "Is time long for you out here?" I tell him it is. For the fly on the wall, time is apt to be long. I awake after midnight for a bit, sweating in the captain's sauna. The ship gives when the rakes hit the dump tables, and the chains rattle like Marley's ghost.

The day dawns bright, the sea blue-black in the low sun. We're getting some rocks but also some big scallops. They might go twenty meats to the pound, John says. This is a Nickerson boat, and John says Nickerson will probably keep her fishing through November. After that comes "rocking-chair money," Canadian government support through the idle winter.

Jean N. is so old she has long ago been written off in the company books. Like most other fishing outfits, Nickerson has a creaking lot of ancients in its fleet, and the trouble will come when they have to be replaced. Even with subsidies, Canadian boats cost more than in the States, sometimes twice as much as the boats built along the Gulf Coast. They may be better for these seas, but not by that factor.

Frank, the skipper, remembers the good, dead days. "Back in 'eighty-one," he says, "the crew wouldn't sit down for dinner. They just had to get their thirty thousand pounds and go home a day early." He once worked trawlers, too, he says. "But the trouble there is that you're gone from home so much. The draggers turn around in port in two days and the scallopers in five. And you never know when you can sleep on a dragger. You just lay down and you got to go again."

After dinner, cleanup begins. The waist is shovelled out, then hosed down, then scrubbed, then hosed down again. Frank hops out to take a hose and show a deckhand how it's done. John works with the men on the starboard rake, feeding new links into a crimper as big as a tree pruner, weaving new net with heavy nylon line, replacing rubber chafing gear. By supper, *Jean N.* is as clean as she can be and keeping an eye on another Riverport skipper heading home, just in case he runs out of fuel. The crew look to themselves as they looked to the boat. Somehow, without showers, they seem showered. They gather in the wheelhouse and smoke — everyone smokes — and hunt along the radio bands for rock music.

Not a bit of light shows at five in the morning. Just the dock lamps and Venus riding close over the stern. The boat we've been escorting decides he has enough fuel left to beat us in, which he does, surging past and in front of us in the last hundred yards. Catcalls on the radio. Frank maneuvers and squeezes in behind, tying up outboard of a stern trawler. Then he raises his hands off the old wheel and brings them down in a hard pat.

"Thank you, Jeannie," he says, "for another one."

Sandy MacLean, the lanky deputy minister of Nova Scotia's fisheries department, told me when I first met him in the summer of 1982 that he had made his own map of the boundary the World Court would eventually draw across Georges Bank. "I have split the intersects all the way down," he said to me. "That's my best guess." Sandy wasn't alone in his guess. A good many Americans and Canadians familar with past decisions of the World Court would be inclined to bet the same way. But Sandy was a bit more emphatic. "I have it dated and signed," he said, "and I think I even have a gold signature on it."

On October 12, 1984, the special panel of the International Court of Justice charged with the Georges Bank case pretty much followed Sandy MacLean's gold-plated hunch. Four of the judges, from Canada, the United States, Italy, and West Germany, voted to split the difference between the expanded Canadian line (the one that ignores the significance of Cape Cod) and the original American line through the channel separating Georges from the Nova Scotian continental shelf. Canada ended up with the tip of the tip of the thumb, including the scallop grounds around Northeast Peak. The four judges did not pay much attention to Canadian claims of economic dependence on Georges or American insistence that they were there first; they based their decision on the geography of the matter, who owned how much coastline along the Gulf of Maine, that sort of thing. The fifth judge, from France, objected, saying he found little basis in international law for the decision. The American and Canadian governments said they would live with the new line and began the prickly business of getting their fishermen to do likewise.

Before the decision, fishing captains of both countries talked in terms of vague fear. Some said they were used to working in the disputed area, that a change — that dreaded element in a fisherman's life — could only be for the worse. After the decision, the fear had something more solid to work on. New Bedford scallopers said that half their grounds were being transferred to Canada. Lobstermen up and down the New England coast said they had been robbed of some of the area's best spots around the northeastern canyon heads, and draggermen worried that an important spawning ground for haddock had been transferred to Canada. Gloucester fishermen talked of organizing a protest sail of boats up to Boston, where a fishing industry exposition was under way, but Angela Sanfilippo of the Gloucester Fishermen's Wives Association was against it. "We don't want the Canadians to feel we're at their mercy," she told me. "And besides," in a sad voice, "they're hurting just like we are."

Up in Lunenburg, Nova Scotia, scallopers found themselves confined to an even smaller piece of the Bank than they had worked in the disputed area. That could only mean more pressure on the stocks or, if fisheries managers like Sandy MacLean had their way, fewer boats fishing. But the more pragmatic in the Nova Scotian outports admitted quietly that it could have been worse. Their country acquired a quarter of the Bank rather than the third or more it had claimed, yes, but what remained was more productive, acre for acre, than what the Americans got. We ended up, they said, with most of the herring, a large share of the lobster, and half the scallops. And besides, they asked, where would we be if the United States had won its claim to the whole show?

Lloyd Cutler still laments the agreement he worked out with the Canadians in 1979 only to see it sunk by American fishing interests. He thinks something similar, some form of catch allocation mechanism, still makes the best sense for the nations now staring at each other across the new Georges line. What he and his Canadian counterpart had in mind "was too sophisticated for the

fishermen," he told me when I phoned him after the
World Court decision, "but perhaps the governments will
step in now." At least one Canadian, a man who helped
engineer what Ottawa regards as its victory at the Hague,
agrees that the Cutler-Cadieux approach would have
been preferable to the World Court decision and should
be resurrected. "Since we are condemned to live together," he told me, "we ought to have the good sense to
be able to share the resources."

Allen Peterson, down at the Fisheries laboratory in
Woods Hole, isn't all that sure there'll be much sharing.
It is going to take time, he thinks, for each government to
confer with its fishermen and much more time to negotiate a joint fishery management agreement for the trans-
boundary stocks — those species, like herring, that swim
back and forth across the new line in finny innocence.
"We may never get to that," Allen said, shortly after the
coordinates of the new boundary were published. "This
is going to be a case of people being asked to give up
something without getting anything directly in return. If
we want access to Canadian groundfish, say, we're going
to have to give them some of our scallops. And how are
the scallopers in New Bedford going to react to that?"
Allen's hope is that the two countries can avoid a standoff
and move toward a neutral Georges Bank commission
that can at least assess stocks and set the catch quotas
necessary to protect them.

Whatever happens, the game will have a different dimension from now on. The Georges Bank decision was
the first in which the World Court dealt simultaneously
with both fishing and seabed rights. That may mean that
the ruling, and its acceptance by two of the most developed nations, can be of real help in settling the scores of
other maritime boundary disputes around the globe. But
it also means that the new line across Georges is meant
for oil as well as for fish. The Canadians, who have held
the drilling permits they issued on Georges Bank under

moratorium pending the World Court decision, are now presumably free to authorize rigs to move onto their part of the Bank, an area containing some of the more promising oil prospects in the western North Atlantic. The United States, under the same presumption, can now lease blocks on its newly acquired strip of what was once disputed area. It is arguable whether fishermen on both sides, fenced in as never before, will abide further drilling on their grounds. Especially those who agree with Doug Dory of the *Ernest E.* that "if they find oil there, that'll be the end of it."

9

Second Flight

THE Goats are gone, the dear old boat-hulled Albatrosses; it would be nice if just one of those airborne
relics had gone from the Coast Guard to some hidden
bay in the Caribbean. The silhouettes out on the flight
line now are low and racy, everything an airfoil. These
new planes are Falcon jets, a French airframe with
American add-ons, and the Coast Guard is pleased with
them. Lieutenant Bob Boetig says tests have proved the
Falcons are the best platform around for the kind of electronics his service wants on its coastal patrol planes.
They can hit 400 miles an hour at altitude and cruise at
250 and rein in to about 130 to drop gear or rig boats.

Bob Boetig will take the flight this April morning in
1984, a training flight that will cut up to Georges Shoal
and then up the wide corridor between the Canadian and
U.S. boundary claim lines. The case won't be decided for
another five months. "The Canadians fly that one day
and we do it the next," Bob says. He takes me over to a
wall chart marked with the claim lines, the two areas the
National Marine Fisheries Service has closed to protect
spawning groundfish, and two long, narrow corridors for
foreign fishermen — one off Georges, the other running
in deep water from off Rhode Island down to Hatteras.

While I get used to that, he phones in for weather information. The sky was clear as gin at sunrise, but the low that has been hanging around for a week threw an arm over Cape Cod by seven, and the wind is rising. Weather tells Boetig to expect low clouds, rain, and gusts to forty.

A crewman walks me over to a Falcon in the hangar and talks me through what I need to know. Exits here, life rafts here. Life vests hanging on the rack, survival suits in a canvas bag. Oxygen masks have a new wishbone device that slaps the cup over the face in a hurry when you need it. One thing, the crewman says, you better empty out before we go. Gent's room is over there through that door.

A warrant officer is with Boetig now, telling him that the damn computer that guards the engines is so sensitive it is registering vibrations in the left engine where there are none to speak of. Just disregard the warning, he says. Boetig looks at the warrant officer solemnly out of his light eyes, then nods his dark head. We got a camera? he asks his copilot. No. So a crewman hunts one up, a full kit for snapping fishing boats and suspected druggies. The traffic, Boetig says, is still coming north.

Three crewmen — one a specialist in maintaining the suite of sensors aboard the Falcon — are getting their gear together. "Hey, Grumpy!" the youngster who checked me out yells to the sensor specialist. "You comin'?" We follow them out, to the white bullet with the slash of Coast Guard red on the nose.

Inside, the Falcon is surprisingly ample, able to carry nine in some comfort. I am to sit in the jump seat, just behind the pilots. My crewman guide installs me, fits on oxygen, intercom, pats my shoulder. Boetig and Craig Veley, the copilot, stare at instrument panels that belong in the Museum of Modern Art. Green lights gleam, yellow lights, dials from Tiffany, hundreds of lovely indications. Straight ahead is the radar, a small screen, and below it computer buttons that control just about every navigational readout the pilots will need.

Boetig starts the engines. I can barely hear them. We creep out to the runway. The pilots' hands, gloved in flame-proof cloth, reach and flick and twirl. A woman's voice on the radio gives us the go. Boetig's right hand pushes the throttles forward, and the thrust flattens me against my seat. No noise, just force. This is hurling, not flying.

We climb, the Falcon quietly singing, up over the forearm of the Cape and out and into the low. The cloud cover is thin, and we break through quickly into sun and level off, on automatic pilot at 13,000 feet. In a few minutes, Boetig sets the descent to 500 feet. We slope down through the cloud rind, and there she is, black rollers in close order drill, epaulets of spume on their shoulders, the old battleground.

All is quiet and tidy here. Where are the activists? Lars Sovik and *Valkyrie* might very well be out here right now, maybe fishing the Leg again. *Ernest E. Pierce* and the rest of them from Lunenburg should be raking and picking what is left of the scallop beds around Northeast Peak. *Delaware* or *Albatross* usually cut across Georges about this time on their spring stock-assessment cruise. Fish prices have been way up, and I wonder if that means the cod or haddock catches are down. The winter here wasn't too bad, but the spring storms have been horrible.

Up on the left side of the modern-art assemblage, a warning light goes on. Vibrations in the left engine. Boetig looks back at me and smiles. After a while, the computer decides not to be so tender, and the light goes out.

Science is still verifying its story on Georges Bank. The monitoring program has finished its first year, and the results indicate that the eight holes in the Bank didn't have any significant effect on bottom life. Eight holes aren't what John Teal had in mind when he conceived of his hobby horse, a giant perturbation on Georges kicked up by rigs and platforms for science to observe. But the program itself, the design and execution of the experiments, has been praised by such meticulists as Howard Sanders of Woods Hole, who undertook the first long-term study of an oil spill in New England waters. It looks as if intensive sampling on Georges will end in a couple of months, though one or two research cruises will be made during coming years to keep an eye on things. Deep-water monitoring has started further south and will probably move into the North Atlantic. The aim there is the same: to figure out what is going on in the water column before drilling, during drilling, and after drilling.

"Pardon me while I have breakfast," says Veley, and pulls out something that looks like a roll; it is, I am glad to note, not a peanut butter and jelly sandwich. Targets are showing now on the radar screen. Boetig says we'll rig a few and takes us down to two hundred feet to begin reading boat names and boat numbers. He selects a blip about ten miles off, to the right, and guides us in.

But will there be any more drilling? I've been hearing rumors that an unnamed oil company expressed interest in a ninth well and did some seismic work. No rig has showed up on the Bank though. Lease Sale Fifty-two is still under injunction. The government took the case to the First Circuit Court of Appeals in the summer of 1983. Three federal judges, sitting where Judge Mazzone sat beneath the thirteen gold stars in the cool blue courtroom in Boston, upheld Mazzone's ruling. But that doesn't seem to make a great deal of difference to the Department of the Interior. Secretary William Clark, who moved over from the White House when James Watt resigned, is moving ahead with Watt's leasing program, though he is nicer about it, more considerate of how environmentalists and, especially, governors feel about things.

Interior has been talking with Massachusetts about how it wants to conduct Lease Sale Eighty-two, which was supposed to have been held two months ago, in February of 1984. Eighty-two is the first area-wide sale in the North Atlantic, and the area is huge.

"Ten seconds on the right." Boetig's calm voice on the intercom.

The people in Rich Delaney's coastal zone management office are pleased about the new cooperative attitude down at Interior. At the same time, they have lost one of their favorite weapons — the consistency argument, the one that says the Department of the Interior must show that a given lease sale off a state's coast is consistent with that state's coastal zone management program. The Supreme Court has ruled that the time to raise the consistency issue is after the lease sale, not before or during, since the Court does not regard a lease sale in itself as being injurious to any coastal zone. Massachusetts' strategy now is to argue for a fisheries deletion area that goes well beyond anything tried so far — four hundred meters, deep enough to protect not only groundfish but lobsters and red crab out on the slope. Douglas Foy and his lawyers at the Conservation Law Foundation have been talking things over, pressing the state, as usual, to be bold. At Interior, quite a few people feel that no matter what they do to be considerate, the commonwealth will take them to court. Not the best atmosphere for reasoning together.

"Side-trawler," a crewman says on the intercom, and reads the boat's numbers painted on the wheelhouse roof. Someone else gets the name. Boetig punches one of his marvellous buttons, and the computer notes the position.

Litigation. We are a litigious society. A Fordham law professor named Jethro Lieberman has written a book under that title. He says we are not exactly number one in the international category of court-going — we share our passion for the suit with such as England, Canada, Australia, and West Germany. But our trend has been up. Lieberman starts his book with a quote from another student of the phenomenon, Eugene Kennedy, who says the seventies launched the balloon. Judges started running hospitals, prisons, schools. Then, says the expert, "hundreds of other issues became the courts' business as the nation lost faith in other forms of negotiation and gave the courts power they did not seek, without any guarantee that they could exercise it prudently or effectively over a long period of time."

The nation is suffering from the fulfillment of the old Mexican curse: "May your life be filled with lawyers." Chief Justice Burger has been saying that for years. "We may well be on our way," he complains, "to a society overrun by hordes of lawyers, hungry as locusts, and brigades of judges in numbers never before contemplated." Some estimators believe the legal industry will grow from over six hundred thousand practitioners now to over a million in fifteen years.

Cecil Andrus thinks that "litigation is wearing the political process thin, and not just in offshore oil. We've got too damned many lawyers and they're coming out of the law schools by the thousands. They bring frivolous and unneeded lawsuits simply because they don't have anything else to do. The public has to defend against those lawsuits." He says that even when he was secretary, say, around 1980, the solicitor's office in Interior had a payroll of over 250 lawyers: "one of the biggest law firms in the city." Andrus admits he needed them, but "it's frustrating just to see them play lawyer. The public is taking it in the neck with the delay" brought on by litigation.

Boetig banks the Falcon and heads for a far target, flying in and out of rain shrouds.

Those who have gone to court for the fisheries have mixed feelings for the process. Steve Leonard, the assistant attorney general who fought Interior with Doug Foy, says that "litigation is a terrible process." It is, he says, a blunt instrument, acceptable "if your object is to keep them off the Bank entirely." The trouble is that the process doesn't lend itself to flexibility, to adjustment. Steve admits that as an environmentalist he often feels that he should be waving a flag, preserving what can still be preserved. "On the other hand" (that most lawyerly of terms), "if you're talking about trying to balance the need for secure energy supplies against the need to protect an important food source — all of the things that go into the Georges Bank decision — throwing it into the hands of a federal judge seems to be the last way I'd do it. There's no way that a judge can balance out all those things. He's not equipped to."

"Litigation," says Doug Foy, "is probably the least efficient and least effective and least satisfying way to resolve these debates." Unfortunately, it is sometimes the only way to get the government to listen. The Georges Bank fight, he thinks, "should have been resolved by sensible interaction among the agencies and the interest groups and by a system for really, seriously protecting that fishery — restricting the sale of rights that came out of those leases, and modifying the whole leasing program in such a way that you had continuing management control of the fishery." Doug, who spends half of his professional (as opposed to managerial) hours in litigation, thinks "the irony is that in exchange for refusing to consider these kinds of issues, the government simply gets endless litigation. They win some, lose some. All over the country, you have different states and different situations. And litigation is not the way to get at them. All it does, if you win, is force the government to do something it doesn't want to do."

Boetig's voice: "Fifteen seconds, on the right."
Crewman's voice: "That's Ocean Spray.*" An old trawler. I think I catch* New Bedford *under the faded name on the stern. Boetig says wooden hulls make poor radar targets, particularly when they're stern to.*

Just after Lease Sale Forty-two, Doug Foy and John Teal and a couple of fishermen and executives from Mobil and Exxon and some others tried something new to Georges Bank. They were invited by a former Boston heating oil salesman turned mediator to see if they could agree on anything about the place. After some waving of battle flags, they did work out agreements, principally on the need to monitor drilling operations — for different reasons, of course. And they put some pressure on Interior to listen to them. Modest enough results, but the extent of their agreement impressed some in the group. Doug says the process had its moments, but even in mediation he thinks in terms of force. "The whole thing depends on the effectiveness of the adversaries in the battle. Mediation works best when you have tough people on all sides and a good mediator. You just give up some stuff and you come into the middle ground."

Mediation or consensus resolution or what the big law firms call "alternative dispute resolution" works better on a regional than on a national basis, Doug thinks, and particularly well in community-conscious New England. "I have for years believed that the real problem on Georges Bank was the federal government. Everyone involved in New England has had a terrible time persuading the agencies that they should be doing something here that was more sensitive and more rational than anywhere else. That's because of the typical bureaucratic attitude that business should be conducted in exactly the same way straight across the country." And because in Washington "they think they're the center of the universe. So you get more litigation there than you do at the regional level." With different words — and, of course, for different reasons — a good many oil men, fishermen, and scientists agree with him.

Ah, the fate of the regulator! Even Garrett Hardin of the tragic commons was against the regulator. In our administration of public resources, Hardin saw a primitive management system overlaid by a veneer of technological fixes. "The laws of our society follow the pattern of ancient ethics," he wrote in *Science* (December 13, 1968), "and therefore are poorly suited to governing a complex, crowded, changeable world. Our epicyclic solution is to augment statutory law with administrative law." We do this because it is impossible to legislate behavior where the customs and tenets of private property don't apply. We delegate the details to bureaucrats and they create regulations. Hardin didn't like what he thought that created: a government by men rather than by law. *"Quis custodiet ipsos custodes?"* he asked. Who will watch the watchers?

The patrol is about over. We fly along the edge of one of the closed areas. Boetig says that the computer has set the course so that any blips to the left of the radar screen will be in the area. Not a target shows. So we lift up through the scud and head home, stooping low over a freighter stranded at Nauset on the outer beach of the Cape by a horrible March storm.

Who will watch the watchers? Doug Foy certainly will, and John Farrington and John Teal of Woods Hole and the Georges men flocking to the fishery council meetings. Angela Sanfilippo of Gloucester will watch, and Lena. The Canadians will be looking south. The oil men will watch and take their actions if the rope of regulations chafes. All believe in, some crave, cooperation. The first time I met Bob Graves of Mobil, he talked about the adversarial mania in the country and how much better it

would be to negotiate. But interests and grievances run
too deep. Jethro Lieberman, the author of *The Litigious
Society*, says industrialization has carried our society to a
point of unreality, where we have forgotten what we are.
Our comparative wealth swaddles us so that we ignore
our limits and nature's. "We cannot see, or do not wish to
admit," Lieberman writes, "that no society can ever rid
itself of everything that could conceivably be called ill-
being, because human beings remain subject to an often
perverse and parsimonious nature, which imposes limits
on everything at which we aim. That we filter our lives
through social and political institutions to shield us from
the worst of nature 'red in tooth and claw,' cannot alter a
fundamental aspect of any world: that resources are always scarce because men and women will always be capable of wanting more than is possible from all the
arrangements in society at one time. . . . In a society that
takes for granted what to any other age would be considered beyond utopia, each harm, every source of ill-being,
cries out for redress."

*Bob Boetig calls the tower at Otis. He and Veley are
going to practice landings. We slide in toward the field,
leaning a little into the crosswind. Boetig goes first,
lands like a leaf, hits the throttles, and rockets up. I
rearrange my stomach and wonder. How many times
will they touch, will they go?*

And how many times will they touch and go over
Georges Bank? It takes trust to stay out of court, and
trust in New England is as scarce a resource as it is anywhere else. We have made things too complex for trust-
ing. So how long? Frank Basile said to me after the
cancellation of Lease Sale Fifty-two that things would
stay as they were until the presence or absence of oil
under the North Atlantic is known.

How long? The combatants are prisoners of their own

combat. "No one wins forever in the offshore," Doug Foy says. Perhaps eventually there will be compromise, coming into the middle ground. But Jethro Lieberman doesn't think so. "Until the day when our institutions can be trusted to serve us as fiduciaries and when we can be educated to understand the limitations of the world we have created, litigation will remain the hallmark of a free and just society."

The Unbidden Bidder

"WELL," said Judge A. David Mazzone on September 20, 1984, "I've tried, in preparation for your hearing, to do as much as I could, which is simply not enough." The judge had agreed to interrupt an important case already in progress to hear the commonwealth of Massachusetts and the Conservation Law Foundation of New England in their arguments against Lease Sale Eighty-two. Now he looked down on Douglas Irving Foy and Steve Leonard and counsel for the federal government and for the oil companies intervening in the case and said, "I'm not sure I can give you the time schedule that you want." So began round four of the fight for Georges Bank.

Sale Eighty-two was announced in the fall of 1982. It started out covering a huge amount of ocean bottom, about 25,000,000 acres of it, from way south of Long Island to the far tip of Georges, from depths of 16 meters to depths of 3,000, and gradually shrank to around 16,000,-000 acres. It was a full-bore James Watt special, carried forward by Watt's successor, William Clark. The Department of the Interior asked oil companies to indicate comparative interest in over four thousand blocks, including acreage in areas disputed by Canada. The department wasn't saying that it would actually lease the disputed bottoms. The idea, following the Watt philosophy, was to give each corporate exploration strategy as extensive a range of ocean to work in as possible. After all, the World Court might accept America's claim. (It did not.) When Interior found it could not await the court's decision and meet its leasing schedules, Secretary Clark divided Sale Eighty-two in half. Part two would be deferred; it would include oily-looking tracts in whatever territory the World Court gave to the United States plus tracts on biologically productive bottoms near the crest and around the canyon heads of Georges Bank which had been put under a year's leasing moratorium by a Congress worried by Interior's fast-track leasing pace. After all, that leasing ban would expire on the first of October, 1984, and Congress might not authorize another. (It did.) As for the 1,138 tracts left in part one, Secretary Clark announced he would open bids on September 26, 1984.

Not if we have anything to say about it, cried the commonwealth of Massachusetts and the Conservation Law Foundation of New England.

They did.

Judge Mazzone had heard most of the arguments before, in the hearing on Sale Fifty-two, which, though cancelled by Interior, now seemed born again as part of the new leasing. What was new, of course, was the size of Eighty-two. Doug Foy and his colleagues argued that no environmental impact statement for a stretch of sea that big could be of much help to the Secretary of the Interior in his appointed task of balancing the goods and bads of the sale, and Mazzone agreed. "This enormous proposed lease area," the judge wrote in his ruling, "completely subsumes prior sales Forty-two and Fifty-two. Those earlier sales form only a *small fraction* of the area the secretary now intends to lease under the cover of only one environmental impact statement. . . . How can the De-

294 Oil and Water

partment of the Interior accurately estimate the impacts of exploration on this massive area in only one document, which is not much longer than those prepared for the prior sales? How can it provide adequate and sensitive alternatives for areas subject to both a Congressional moratorium and the vagaries of international litigation? Most important, how can this court rest assured that the specific and harmful impacts that may be felt by the most admittedly delicate part of this huge tract will receive adequate and separate analysis and protection when it is buried or 'diluted' in such an enormous plan?" Citing these apprehensions and what he considered other shortcomings in Secretary Clark's handling of the sale, Mazzone ruled in favor of Massachusetts and the Conservation Law Foundation, and the Gloucester Fishermen's Wives Association, and the Sierra Club and Greenpeace and the other plaintiffs.

Farce followed. The government lawyers acted at first as if they would appeal Mazzone's preliminary injunction. But then Interior cancelled Sale Eighty-two. For the first time in federal offshore oil leasing, not one single bid had been received by deadline time the day before the scheduled sale date. Well, not one single bid from an oil company. The one organization that did submit bids was an outfit more unwelcome around the public affairs offices of Interior than news of a dozen dry holes in a row: Greenpeace, they of the rubber rafts and the signs reading "Oil and Water Don't Mix," and, as has just been noted, coplaintiffs along with Doug Foy in the battle for Georges Bank. The rogue bids were, of course, symbolic in intent and, undoubtedly, in amount. "We have been saying that Georges Bank is priceless and unquantifiable," Greenpeace announced, "and our bids reflected our beliefs." But not even the truest of the organization's believers could have imagined how much publicity for their cause those unbidden bids would generate. "I can't think of a worse loss for Interior," Foy told me a few weeks after Mazzone's decision. "They lost the lawsuit, which is bad enough. Then nobody showed up for their party, which is twice as bad." He paused and threw his head back and laughed. "Except Greenpeace," he said, chortling at the ceiling, "which is three times as bad."

In Interior's defense, it should be said that the agency cannot control the moods and stratagems of the oil companies. Several companies did support Sale Eighty-two, at least in the early stages of its multimillion-dollar development process. The Interior spokesperson I talked with said that perhaps the companies lost interest because of all the territory deleted from the sale, in large part to meet objections from states like Massachusetts, from the Congress, and from federal agencies like the Coast Guard (which worries, among other things, about putting rigs and platforms too close to shipping lanes). But I remembered Bob Graves of Mobil saying back when Eighty-two was just a babe that he didn't have many darts left to throw at Georges Bank. Eight dry holes on the Bank and the generally subdued outlook for the world oil market since then could have reduced those bidding darts to zero just as easily as the tract deletion. Nonetheless, Georges Bank, either on its own or as part of the national offshore leasing program, retained sufficient importance so that oil companies were present as intervenors on the government's side during round four.

Judge Mazzone tacked an addendum onto his ruling. He said that about five hours after he had issued his preliminary injunction restraining Secretary Clark from conducting Eighty-two, part one, the secretary had asked for another hearing. This time, his lawyers argued that since the sale had been cancelled and because there were no bids from the oil companies, Mazzone should deny the motion for a preliminary injunction; the issues, they said, were now moot. Shades of Judge Garrity and that wonderful business at the end of the Sale Forty-two hearing: "How can I grant a stay of my order to stay the secretary's action?" How, indeed! A. David Mazzone, having slaved away on yet another last-minute hearing on federal efforts to lease Georges Bank, declined to throw all that work into the wastebasket. He did so, he said in his addendum, "first, because my order had issued. Also, only part one of Sale Eighty-two was cancelled; the future as to part two is uncertain."

As is, of course, the future of the great thumb itself and the wild waters over it. The Department of the Interior has since cancelled part two of Eighty-two, and no more sales will be held on or near Georges Bank for several years. But the fight isn't over yet.

Index